BATH SCIENCE 5-16

EARTH

Nelson

Thomas Nelson and Sons Ltd
Nelson House Mayfield Road
Walton-on-Thames Surrey
KT12 5PL UK

51 York Place
Edinburgh
EH1 3JD UK

Thomas Nelson (Hong Kong) Ltd
Toppan Building 10/F
22A Westlands Road
Quarry Bay Hong Kong

Thomas Nelson Australia
102 Dodds Street
South Melbourne
Victoria 3205 Australia

Nelson Canada
1120 Birchmount Road
Scarborough Ontario
M1K 5G4 Canada

© Thomas Nelson and Sons Ltd 1993

First published by Thomas Nelson and Sons Ltd 1993

ISBN 0-17-438435-1
NPN 9 8 7 6 5 4 3 2 1

All rights reserved. No paragraph of this publication may be reproduced, copied or transmitted save with written permission or in accordance with the provisions of the Copyright, Design and Patents Act 1988, or under the terms of any licence permitting limited copying issued by the Copyright Licensing Agency, 90 Tottenham Court Road, London W1P 9HE.

Any person who does any unauthorised act in relation to this publication may be liable to criminal prosecution and civil claims for damages.

Designed by Fox and Partners, Bath

Printed in Hong Kong

The Project Team
The *Bath Science* Project Team is directed by Professor Jeff Thompson CBE and Dr Martin Hollins of the School of Education, University of Bath. The authors for the Key Stage 4 materials were:

John Collins	John Harris	Christine Harrison	Martin Hollins
Theresa Kinnison	Janet Major	David Sang	

with editorial supervision by Janet Major, Christine Harrison and Martin Hollins. *Earth* was written by Janet Major, Christine Harrison, John Collins, Martin Hollins and John Harris.

Acknowledgements
The authors and publishers wish to acknowledge with thanks, the following photographic sources: Aerofilms Ltd p. 47; Barnaby's Picture Library pp. 29, 59, 60 top and bottom; J. Allan Cash Photolibrary pp. 45 left, 46, 51 right, 53 right, 59 top, 59 bottom, 61 bottom; Martin Collins p. 97 top; Mary Evans Picture Library pp. 12, 79 left, 101 top, 108, 109; Fitzwilliam Museum, Cambridge p. 125 left; FLPA pp. 6 left, 14 top right, 33, 34 left, 34 top right, 36; GeoScience Features pp. 64, 76, 83; Sheila Goater p. 64; The Guardian p. 29; Landforms pp. 6 right, 62, 63 left, 63 right, 67 centre, 68 top, 75 top, 75 bottom; The Mansell Collection pages 106, 107; National Meteorological Office p. 27, 29, 31; National Radio Astronomy Observatory p. 125 right; NHPA pp. 13, 34 bottom right, 35, 37 right, 41, 50; Oxford Scientific Films pages 5 right, 7, 61 right, 65, 66 top, 72 centre left, 72 top; Picturepoint Ltd pages 4 bottom right, 5 left, 10 top right, 14 bottom, 21, 37 left, 38 left, 45 right; Popperfoto p. 99; Valerie Randall pp. 4 top right, 10 left, 10 bottom, 44, 60, 61 bottom right, 96; Ann Ronan Picture Library pages 11, 29, 85, 97 bottom, 126; Science Photo Library pp. 4 bottom left, 8, 19, 20, 22, 28, 57, 58 centre, 62, 67 bottom, 68 bottom, 74, 80, 93, 98, 101 bottom, 102, 104, 112, 115, 117, 118, 119; Science Museum p. 30; Shell UK Limited p. 82; Tony Stone Worldwide contents page, 4, 38 right, 49, 60 left; Alan Thomas pp. 66 bottom, 67 left, 67 right, 72 right, 72 centre left, 79; Tropix pp. 42, 43, 53 left; Zefa Picture Library p. 67 bottom.

Illustrations drawn by: Tek Art pp. 5 left, 10, 11, 12, 13, 15, 16, 17 bottom right, 18, 22, 25, 26 top, 27, 28, 34, 35, 38, 39 left, 40 bottom, 40 top, 49, 51, 52 top, 53 top, 54 left and right, 55 top centre, 62 bottom right, 64, 66, 69, 70, 71 top right and bottom, 72, 73, 74, 76, 77, 78, 79, 80 bottom, 81 top and bottom, 82, 83, 84, 85, 87, 88, 90, 91, 92, 93, 94, 95, 96, 99, 100, 102, 103, 104, 105, 110, 111, 113 top, 114, 123 top; Sarah Mabbutt pp. 5 right, 7, 17 left and top, 20, 26 bottom, 36, 40 bottom, 45, 47 left, 81 left, 113 right, 120, 121 right, 123 bottom, 124; Gillian Hunt p. 9; Lynn Williams pp. 14, 54 top left, 56 left, 59, 62 bottom left, 63, 80 top, 86, 117, 118; Maggie Mundy Illustrators' Agency pp. 32, 33, 39 right, 40 top, 47 right, 53 bottom, 62 top left and right, 65; Steve Noon pp. 50, 52 bottom, 55 left, 75, 107, 108, 121 left; Jolyon Webb p. 71 left; David Pow p. 113.

The publishers have made every effort to trace the copyright holders, but where they have failed to do so they will be pleased to make the necessary arrangements at the first opportunity.

GLOBAL DECISIONS

Contents

Unit 1	Air and atmosphere	4
Unit 2	Water in the environment	14
Unit 3	The weather	22
Unit 4	Understanding the environment	32
Unit 5	Land use and management	44
Unit 6	Local investigations	54

UNIT 1
AIR AND ATMOSPHERE

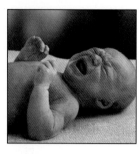

A newborn baby takes its first breath. During its life it will take about seven hundred million breaths

Section 1.1

WHAT IS AIR?

In the sixth century BC the philosopher Anaximenes realised the importance of the air. He thought that air was the foundation for everything in the universe and that what was important to living things was how concentrated the air was. A little later than this air was grouped with three other 'elements': earth, fire and water. It was thought that these four 'elements' were needed to make all known substances.

Air is both invisible and tasteless, and yet it is very important to us. But what exactly is air?

A. Discuss

- Discuss whether each of these statements about air is true or false.

  1. Air is a compound.
  2. The Earth is an unusual planet because the air making up the atmosphere contains oxygen.
  3. The most abundant gas in the air is nitrogen.
  4. The only useful gas in the air is oxygen.
  5. Hydrogen gas can be separated from the air by fractional distillation.
  6. The air making up our atmosphere today has not changed since it was made 2000 million years ago
  7. Air weighs nothing.
  8. Air can be turned into a liquid.
  9. Air is an important raw material for the chemical industry.
  10. On a day with a light breeze, the air you breathe in may have come from 300 kilometres away.

Only in a limited region around the Sun are the temperatures suitable for the existence of life. Too near is too hot; too far is too cold. Within this region are the planets of Venus, Earth and Mars. The atmosphere of Venus is rich in carbon dioxide. This traps most of the heat that is reflected from the surface of the planet and results in the surface of the planet being at about 485°C. The atmosphere of Mars is very thin and does not insulate the planet sufficiently. Temperatures on Mars fluctuate between 16°C during the day and -80°C at night.

Earth is unique in the solar system in that it has the right type of atmosphere to support life as we know it. This atmosphere separates the planet's surface from space.

B. Work out

- What evidence do the three photographs above provide for the existence of an atmosphere around the Earth?

- Give suitable briefs for a photographer to take three more photographs to add to this series. For each one add a note to say how it is providing evidence of the existence of an atmosphere.

4   EARTH : **GLOBAL DECISIONS**

Experiments to discover more about the air were carried out in the 1700s. At that time it was thought that air was an element, but Scheele, in 1777, found out that this could not be true. He thought air was made up of two gases: 'fire air' which we call oxygen and 'foul air' which we call nitrogen. He thought there was one part of oxygen to three parts of nitrogen (by volume). Lavoisier was working on this in France and thought the proportions were one quarter oxygen and three quarters nitrogen. Priestly carried out experiments in which he reacted metals with air. He thought that air contained one fifth oxygen and four fifths nitrogen.

The following apparatus can be used to test these predictions of how much oxygen there is in the air.

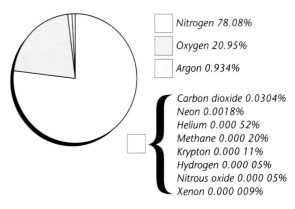

Nitrogen 78.08%
Oxygen 20.95%
Argon 0.934%

Carbon dioxide 0.0304%
Neon 0.0018%
Helium 0.000 52%
Methane 0.000 20%
Krypton 0.000 11%
Hydrogen 0.000 05%
Nitrous oxide 0.000 05%
Xenon 0.000 009%

Average composition of clean, dry air on a volume % basis

Reading	Remaining volume of air (cm³)
1	84
2	79
3	79

Observation: About half the copper was black the other half was a copper colour.

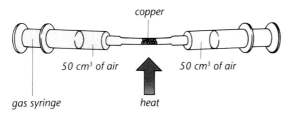

The table on the left shows some typical results for this experiment.

The pie chart shows that air contains a number of different gases. The proportion of these, particularly carbon dioxide and water vapour, varies. Depending on where the sample is taken from, air also contains a varying amount of solid matter.

C. Work out

1. Calculate the percentage of gas left in the apparatus shown in the diagram.

2. What percentage of air has been used during the experiment?

3. Write a conclusion to the experiment. Make sure that you refer to the original predictions.

4. Explain:

    a. why the apparatus was heated and cooled three times before taking the final volume reading;

    b. why the apparatus was allowed to cool before reading each syringe;

    c. how and why the copper would change during the experiment.

5. While carrying out this experiment, you overhear one of the class saying *'This is a complicated experiment; when a burning candle is floated on water under a bell jar, it shows just the same thing.'*

    Is this correct? Write a suitable reply explaining the reasons for your answer.

D. Plan and test

- Choose one of the variable components of air, such as the amount of dirt, to investigate in more detail.

- Think carefully about what you know about the air already, and then make a prediction or hypothesis that you can investigate. (For example, *If there is more traffic there will be more dirt in the air because ...*)

- Work out a plan for investigating your prediction or hypothesis. Check your ideas with your teacher before you begin.

Air contains a varying amount of solid matter!

Section 1.2

LAYERS OF THE ATMOSPHERE

If you have ever been in an aircraft you will have your own evidence that the atmosphere is in layers

This photograph was taken in August, at a height of 3573 metres. The glacier is about 800 metres thick and never completely melts

It is always sunny above the clouds. Aircraft rise through the clouds to reach their cruising height which could be 12 000 metres above the clouds. If you have even been in an aircraft that rises through dense clouds, you will know how it feels. The aircraft gets bumped about a bit, the pilot usually says there is 'slight' turbulence. These clouds seem to cling to Earth in a layer. The atmosphere is a blanket of gases which surrounds Earth. It reaches into space for 800 km (500 miles) but becomes very thin after the first 80 km (50 miles). The atmosphere can be thought of as being in a number of layers. This is quite surprising when we consider how the particles in a gas are always moving about.

It certainly does get cooler as we go higher; approximately a drop in temperature of one degree Celsius for each 150 metres.

However, as we go away from Earth it only gets colder up to a point and then the temperature does not seem to drop any more. At the poles this happens at a height of 8 km, where temperatures as low as -60°C are recorded; at the equator it is noticed at a height of about 15 kilometres. If all these points were plotted around Earth it would be possible to imagine this as a thick layer, although it would be thinner at the poles and thicker at the equator. This layer nearest to Earth is called the **troposphere**, from the Greek word 'tropos' which means a turn. It is in the troposphere where most (about 75%) of the mass of the air is found along with most of the water vapour. Here the air really turns and moves about. It is the layer where we see clouds and this is where the weather exists.

Going outwards from Earth there is a second layer which is twice as thick as the troposphere. This layer reaches up to 50 km from Earth. It is called the **stratosphere**. Stratos is the latin word for something flat, in layers. It is the stratosphere which contains the ozone layer (at about 30 km) which absorbs the solar energy of the Sun. The layer is not very thick, but it generates heat as it absorbs ultraviolet radiation. It also acts as a large umbrella which filters out most of this harmful radiation and prevents it reaching Earth. This explains why the temperature rises from where the stratosphere meets the troposphere to its outer margins from Earth.

> ### E. Discuss
>
> - Without ever going up into the atmosphere it is possible to predict some features of it. For example:
>
>   1. Air must be densest at the surface of Earth.
>
>   2. As we go higher into the first 10 km of the atmosphere it gets cooler.
>
>   Discuss each of these predictions and use your knowledge of gases to give reasons for them. Try and come up with some evidence for the predictions being correct.

There are three more layers of the atmosphere beyond the stratosphere before space is reached. The third layer out from Earth is called the **mesosphere**, this is the layer between 50 and 80 km. The temperature drops gradually down to about -90°C at the outer limit of this layer. The fourth layer out is the **thermosphere**. It is a thick layer, from 80 km to 500 km. Here the pressure of the air is about one millionth of an atmosphere (the normal pressure at the surface of Earth). Even though this layer is so far from Earth, we may see evidence of it if we are in the right place at the right time. This is the Aurora Borealis of the Northern hemisphere (or Aurora Australis in the Southern hemisphere) when coloured lights are seen in the sky. This occurs when streams of electrically charged particles, formed as an outburst from the Sun, ionise the gases of the atmosphere by splitting apart the molecules. The temperature of this layer increases to about 1500°C due to the ultraviolet radiation, gamma rays and X-rays absorbed from the Sun. The mesosphere and thermosphere are sometimes just called the ionosphere.

The fifth and outer layer of the atmosphere, beyond 500 km, is known as the **exosphere**, where only a few atoms of oxygen, helium and hydrogen are found.

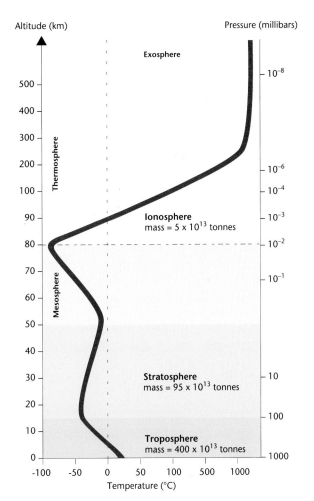

Layers of the atmosphere

F. Work out

- Use the information in this section to complete Worksheet GP1 *The layers of the atmosphere*.

G. Research   EXTENSION

- Find out some height records and add them to your 'layers of the atmosphere' worksheet.
- Sometimes the words 'hydrosphere' and 'biosphere' are used when discussing the atmosphere. Find out what these words mean and add them to your worksheet.

Aurora Borealis

Section 1.3

WHERE DID THE ATMOSPHERE COME FROM?

Soon after the formation of the planets (about 4500 million years ago) the planet Earth may have been very much like Jupiter is now, with a thick layer of gases surrounding the core. It is thought that as each planet was formed it had its own atmosphere of hydrogen and helium, called the primary atmosphere. These two gases still make up 99% of matter in the universe. (You will find out why in the topic *Earth and universe*.) It is thought that, during a period of intense solar activity, radiation from the Sun removed the light gas atmospheres from the planets near to the Sun. The fact that the outermost planets, such as Jupiter and Saturn, still have thick atmospheres of helium and hydrogen around small cores support this idea.

Turbulent cloud formations in Jupiter's atmosphere. The Great Red Spot (top right corner) is believed to be a giant storm system

In place of this early atmosphere another atmosphere developed on Earth. This was formed by Earth itself. The surface of Earth was molten for many millions of years after it was formed. Eventually a thin crust formed on the outside but even then there were many volcanoes which erupted. These gave off a mixture of gases, very similar to the gases given off by volcanoes nowadays. Large quantities of methane, nitrogen, ammonia, carbon monoxide and possibly carbon dioxide were released. There was little trace of any oxygen and so the mixture would have been poisonous to living organisms as we know them today. It was this mixture of gases from the volcanoes which formed the secondary atmosphere of Earth.

Earth gradually cooled down. When it had cooled below 100°C the water vapour in this atmosphere condensed and fell as rain. This happened about 3800 million years ago when the first oceans are thought to have appeared. The water formed lakes and seas in the hollows on the surface of Earth. Conditions would have been just about right in these pools for early forms of life. At a depth of about 10 metres, most of the ultraviolet light from the Sun would have been filtered out. However sufficient ultraviolet light would remain to encourage the type of chemical reactions that scientists think led to life. About 3000 million years ago, the first bacteria and primitive plants appeared. These plants were probably algae. They used the light from the Sun to produce their own food by photosynthesis. As they did so, they released oxygen into the atmosphere as a waste product.

Photosynthesis
carbon dioxide + water ⟶ carbohydrate + oxygen

Once oxygen molecules were released into the atmosphere, ultraviolet radiation from the Sun broke down some of the molecules into oxygen atoms. These would then have combined with the oxygen molecules to form ozone.

Ozone itself is very unstable and would have turned back into oxygen molecules and single oxygen atoms.

About 400 million years ago the amount of ultraviolet radiation had been reduced sufficiently for the first land plants to grow on Earth.

As the number of plants increased on Earth so the amount of oxygen and ozone would have increased also. This provided an important layer protecting Earth from harmful ultraviolet radiation. Beneath this layer was an atmosphere rich in oxygen. Oxygen is reactive and organisms adapted to use this gas. By combining oxygen with carbon from their food, energy could be released (as in respiration in humans). This energy could be used for different functions. Hence simple animals developed which did not rely directly on sunlight for their energy as did the first plants.

The earliest land animals were a type of lung fish

This led to a great increase in the amount of oxygen in the atmosphere and the first land animals, crossopterygians, a type of fish with lungs went onto land 350 million years ago.

H. Work out

- Add the main events in the origin of the atmosphere to the graph on Worksheet GD3 *The developing atmosphere*.
- The graph shows the amount of oxygen in the atmosphere as a percentage of present day levels. Calculate the actual percentages for A, B and C.
- Imagine that you were giving a talk on this topic. Prepare a design brief for a set of slides that you could use. It may help to plan this on a table such as below. (An example has been done for you.)

Detail of slide	Brief notes of what I will say as slide is being shown
1 Slide of Jupiter showing whole planet.	When Earth was formed, 4500 million years ago, it was much like Jupiter – thick atmosphere.

The gravity of Earth is sufficient to maintain the atmosphere in place. Even though the atmosphere moves in a complex way and its vertical structure is also complex the chemical composition of the atmosphere is remarkably constant up to a height of 16 km. If a sample of dry, unpolluted air was taken from anywhere in this blanket and analysed the proportion of the major components would be very similar to that shown in the pie chart on page 5. A number of factors act together to maintain this atmosphere, for example volcanoes, chemical reactions (in particular in photosynthesis and respiration), radioactive decay and human activity.

There is however, now evidence that this natural balance is being altered by human activity. One example of this has resulted in the fall in concentration of the ozone – the so called 'hole in the ozone layer'.

Ozone is an important gas in the stratosphere. It is a form of oxygen which has three atoms in each molecule (O_3). Ozone molecules are being made and broken apart all the time in the stratosphere.

Making ozone

$$O_2 \xrightarrow{\text{ultraviolet radiation}} O + O$$

oxygen molecules → free oxygen atoms

$$O + O_2 \longrightarrow O_3$$

ozone molecules

(This reaction only happens if another molecule, usually nitrogen, is present to take up the energy released in the reaction.)

Breaking up ozone

$$O_3 \xrightarrow{\text{ultraviolet radiation}} O + O_2$$

Usually, as ozone is broken apart to form free oxygen atoms, these react with oxygen to form more ozone. This means that the composition of the gasses in the stratosphere are in balance.

The ozone hole above Antartica. In 1987, at its biggest, it covered an area the size of the United States

The ozone hole was first noticed in Antarctica in 1982, and was later confirmed by satellite data. Each year between September and October the amount of ozone in the stratosphere is depleted. This is the period of the Antarctic spring, and as the Antarctic summer develops the ozone concentration increases. Scientists are trying to discover the reason for this cycle. There appear to be two important causes, the use of chlorofluorocarbons and the exhausts from supersonic aircraft.

1 Use of chlorofluorocarbons

These gases are used in refrigerators (in the cooling system), as cleansers in the electronics industry and to make bubbles in plastic foam. They were also the main propellants in aerosol sprays, such as hair sprays, until an international agreement in 1987 restricted this use.

In 1986 each household in Britain used, on average, 25 aerosols

Surprisingly these chlorofluorocarbons (CFCs) are such a threat to the ozone in the stratosphere because they are unreactive. They are non-toxic, but due to their stability they spread through the atmosphere and some are carried up to the stratosphere. Here these molecules are broken up and release chlorine atoms. The chlorine atoms react with ozone breaking it down to ordinary oxygen:

$$Cl + O_3 \longrightarrow ClO + O_2$$
$$ClO + O \longrightarrow Cl + O_2$$

Chlorine atoms are produced in these reactions and so continue to be present to destroy the ozone. It is the ozone in the stratosphere that absorbs much of the ultraviolet radiation from the Sun. Some ultraviolet radiation does however reach Earth – it is responsible for the tanning action on fair skin. It also causes skin cancer and eye disorders such as cataracts.

Concern over the reduction of the ozone layer is based on research data. For example, the United States Environmental Protection Agency calculates that for a 1% decrease in the ozone concentration of the stratosphere there will be a 5% increase of non-malignant skin cancers. At present radiation, known as UV-C (ultraviolet radiation of wavelength 240-290 nanometres) does not reach Earth. Might it do so in the future? What would be the effects of this? In the laboratory it has been found to destroy RNA, DNA and proteins.

2 Supersonic aircraft

Supersonic aircraft fly in the stratosphere. There is little mixing of this layer with the troposphere below and so the oxides of nitrogen (NO and NO_2) that the aircraft release in the exhaust gases remain in the stratosphere, possibly for years.

$$NO + O_3 \longrightarrow NO_2 + O_2$$
$$NO_2 + O \longrightarrow NO + O_2$$

Notice that NO and NO_2 are produced as well as being used up in these reactions and so continue to be present to destroy the ozone.

Ultraviolet radiation may produce a pleasing sun tan but it can be harmful to the skin

Sun tan lotions can help to protect the skin from ultraviolet radiation

---

I. Think about **EXTENSION**

- Imagine that you have been asked to write a revision sheet on 'The hole in the ozone layer'. The best one may be copied for each member of the class. The revision sheet must be concise and include accurate scientific facts. If possible, include one or two activities (such as a crossword) that may help students.

Section 1.4

THE ATMOSPHERE AS A RESOURCE

We are surrounded by a very important mixture – air! There seems to be an endless supply in the atmosphere. Air is a major raw material of the present chemical industry and, not surprisingly, was used in one of the earliest chemical industries in which iron was made in the Middle East more than 4000 years ago. Here the air was added in short blasts from bellows to enable the temperature in the simple clay furnaces (called bloomeries) to reach the required temperatures of up to 1500°C.

There are many different types of iron and it can be made into many different shapes. Iron and the iron alloy steel were important materials needed to make the machines, buildings and railways of the industrial revolution.

In the 1850s the iron and steel industry of Great Britain was prosperous and probably the most technologically advanced in the world. Items made from these materials were found in the workplace, the street and the home.

Thomas Bradford's rotary washing machine for large families (ca 1895)

In the 1860s, Henry Bessemer developed what became known as the Bessemer Process. This had a major impact on the whole industry and enabled large quantities of metal to be made simply and cheaply. The process was very fast. It involved an exothermic reaction and so it needed no fuel to maintain it. Modifications of the process which occurred during the next twenty years enabled good steel to be made out of virtually any iron. The present day 'basic oxygen' process, which is the major process for making steel, is a variation on this.

This is the major process for making steel. The furnaces can take up to 350 tonnes of charge and turn it into steel in less than 40 minutes

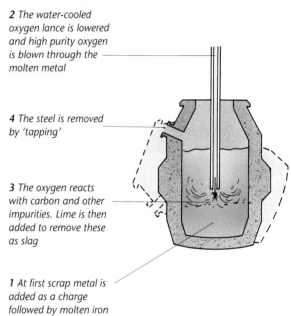

2 The water-cooled oxygen lance is lowered and high purity oxygen is blown through the molten metal

4 The steel is removed by 'tapping'

3 The oxygen reacts with carbon and other impurities. Lime is then added to remove these as slag

1 At first scrap metal is added as a charge followed by molten iron

The 'basic oxygen' process for making steel

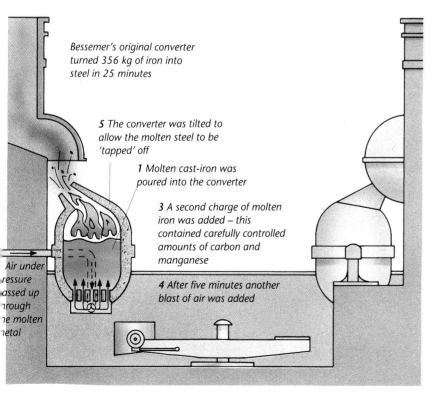

Bessemer's original converter turned 356 kg of iron into steel in 25 minutes

5 The converter was tilted to allow the molten steel to be 'tapped' off

1 Molten cast-iron was poured into the converter

3 A second charge of molten iron was added – this contained carefully controlled amounts of carbon and manganese

4 After five minutes another blast of air was added

Air under pressure passed up through the molten metal

A pair of converters in the Bessemer process for making steel

J. Research

- Compare the Bessemer and 'basic oxygen' processes for making steel. It may be helpful to draw up a table of similarities and differences.

- Henry Bessemer developed his process in the 1860s. Make a list of 10 steel items that we use today that would not have existed in the nineteenth century. For each one say what item would have been used instead.

EARTH: **GLOBAL DECISIONS**

Nowadays the steel industry requires pure oxygen. Pure oxygen is also needed for hospitals, diving, space travel and industries requiring metals to be cut. This oxygen is taken from the air by a process called fractional distillation. The air is liquefied and as it is a mixture of gases these gases can be extracted as it is heated.

1 Filter and drier remove dust, water vapour and carbon dioxide

2 Air is compressed

3 Liquid nitrogen cools the air

4 The air is allowed to expand rapidly. As it does so it cools even more and turns into a liquid

5 Liquid air is a mixture. The components of the air boil at different temperatures and are separated by fractional distillation

Fractional distillation of liquid air

K. Research **EXTENSION**

- The table on the right gives the boiling point of the gasses present in liquid air in the distillation column:

  What is the highest possible temperature for the air still to be liquid at the base of the distillation column? Give a reason for your answer.

- The distillation column is coolest at the top. List the gases in the order they would be removed from the bottom of the column.

- Carbon dioxide does not behave as you would expect when it is cooled. Find out what happens when it is cooled and use this to explain why it is removed at the first stage of the process.

- Explain why it is important to also remove dust and water vapour during this first stage.

- Suggest why liquid nitrogen is used to cool the air during the third stage of the process.

- Find out an important use for each gas produced by liquefying air and give a reason for its use.

Gas	Boiling point (°C)
argon	-186
helium	-269
krypton	-152
neon	-246
nitrogen	-196
oxygen	-183
radon	-62
xenon	-107

Together with iron, coal was the other vital factor for the industrial revolution. Without this there would not have been sufficient iron to meet the demand, and steam power would not have been available for the factories and necessary transport. The steam engine was really the first man-made source of power and replaced the reliance of industry on wind, water and muscle power. By 1880, 154 million tonnes of coal were produced, an increase of over 600% in 50 years.

A Sheffield landscape in 1885

Coal is made due to the action of bacterial decay, pressure, temperature and moisture on vegetable matter over a very long period of time. Chemically coal is mainly carbon, and when it is burnt it forms mainly carbon dioxide and water. However, other products may be produced which can be harmful, such as sulphur dioxide and carbon monoxide.

If the air supply to the burning coal is poor, carbon monoxide and soot may form, instead of carbon dioxide.

In a plentiful supply of air:
$C + O_2 \longrightarrow CO_2$

In a poor supply of air:
$3C + O_2 \longrightarrow C + 2CO$

Carbon monoxide is poisonous but it is the soot which people notice. John Evelyn voiced his concern about the soot from burning coal as long ago as 1661.

It is this horrid Smoak which obscures our Churches, and makes our palaces look old, which fouls our Cloths and corrupts the Waters .... It is this which scatters and strews about those black and smutty Atoms upon all things Where it comes, insinuating itself into our very secret Cabinets, And most precious Repositories.

from *Fumifuginm*: or *The Smoake of London Dissipate 1661*

November 1879 saw the beginning of a dense fog which lasted four months, and many residents demanded legislation for smoke abatement, but no laws were passed.

In September 1952 another severe smog hit London. Pedestrians could not see (some fell into the Thames), cars ran into each other and aircraft could not see to land. During the weeks that followed 4000 people died of respiratory disorders. This led the government to publish the Clean Air Acts in 1952 and 1956 which set up 'smokeless zones' where only smokeless fuels could be burnt. Since then the number of hours of sunshine in London has more than doubled. Smog alerts are common in some other cities with heavy traffic. In Mexico City in January 1989 the smog was so bad that children were given the whole month off school.

Smoke is not the only type of pollution produced by coal. An important impurity in coal is sulphur. When this burns it forms sulphur dioxide. This passes into the atmosphere and dissolves in water making an acidic solution, contributing to acid rain. Other major contributors to acid rain are nitrogen oxides, formed also from burning fuels. The pH of pure water is slightly acidic due to CO_2 being naturally absorbed by rain water but it has been found that much of the rain falling over Great Britain is much more acidic and has an average pH below 5. Acid rain attacks metal and stone and can cause severe harm to living organisms as it affects water quality, forests and soils. Water of a low pH dissolves toxic minerals containing aluminium, cadmium and mercury from the soil. These are much less soluble in more neutral solutions and so are 'locked up' in the soil. Acid rain allows these metals to pollute drinking water supplies.

L. Investigate

- Imagine that a small factory is suspected of producing fumes which contribute to acid rain. How would you prove this?

- You are given a number of samples of elements which are burnt during the manufacturing processes at the factory. Find out which ones, if any, could contribute to acid rain.

- Prepare an Inspector's report for the management of the factory.

- Suggest how the factory could reduce the output of acid fumes.

Guideline for sulphur dioxide is 40-60 mg/m³.
Some newspapers, especially in Europe, publish figures daily, e.g. Zurich 5, Basel 9, but Milan would average 180!!

Lichens are particularly sensitive to small changes in pH

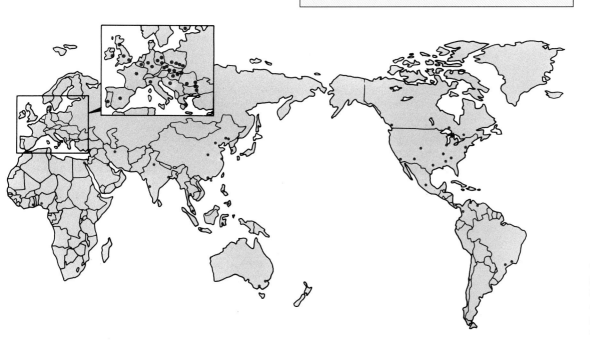

The cities shown by the dots exceed the pollution guidelines for several substances. The air is a clear health hazard, at least for part of the year. Can you identify some of these cities?

UNIT 2
WATER IN THE ENVIRONMENT

Section 2.1
THE GLOBAL WATER BALANCE

Earth from space

Water can produce spectacular effects on the land

In Great Britain we frequently experience part of the system of water circulation

From space the Earth is an attractive planet; brown patches and blue oceans can be seen, even if at times they are hidden by swirling clouds in the atmosphere. Of all the planets the Earth is unique in having liquid water. The Earth is the only planet where the conditions of temperature and pressure are such that water exists in the three forms of ice, liquid and vapour. It is the water in the atmosphere, oceans and ice caps that is largely responsible for many aspects of the world's climate and also the variety of life on the Earth.

There are two main theories on the origin of the oceans. One states that as the Earth solidified water was probably produced by volcanic eruptions. The alternative theory considers that all the water in the oceans has resulted from the crystallisation of the granite continental crust. The water of the planet has been very important in forming the structure of the surface of the Earth as we know it today.

The whole system of water circulation is known as the world hydrological cycle. You will already know the basics of this system as the water cycle, but you may not have considered how the amounts of water involved in each stage compare.

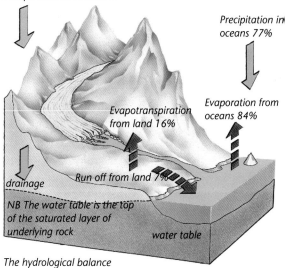

The hydrological balance

The surface of Earth gains water mainly through precipitation. This is usually rain or, in high mountain areas, snow. Quite a lot of water also arrives from the atmosphere as fog or dew. The main losses are from evaporation and transpiration. Water gain and water loss are linked by the flow of water in rivers and oceans, with water being stored in the oceans, ice-caps and as underground water. It is difficult to be precise about the amount of precipitation in different parts of the world as data on rainfall is not available everywhere.

Data on evaporation is even more difficult to obtain. We cannot see evaporation but only the evidence of it, for example in the drying up of a puddle. The other important way in which water is returned to the atmosphere is by transpiration from plants. It is frequently difficult to separate these two processes and so the composite term evapotranspiration is used. In linking up the input of water and the output of water a lot happens. Water moves in the atmosphere, through the land and in the oceans. The routes that the water takes when it reaches the ground depends to a great extent upon the area in which the precipitation falls.

EARTH : **GLOBAL DECISIONS**

A. Work out

- Draw a sketch to summarise the main factors in the hydrological cycle.
- The ways in which the surface of Earth gains water can be thought of as *inputs* and the ways in which water is lost from Earth as *outputs*. Add to your sketch:

  the main inputs in red;

  the main outputs in blue;

  the main linking systems in green.

  Make sure that you include the percentages of each of these.

- Label the main stores of water in black.
- What are the most important factors influencing high rates of:

  precipitation;

  evapotranspiration?

- Predict which areas of the Earth will have:

  high precipitation rates;

  high evapotranspiration rates.

When considering any one point on Earth, the input of water is unlikely to balance the output. Over the oceans, evaporation is high and precipitation is relatively low and so the atmosphere gains more moisture than it loses. Over continents, evaporation is usually less than precipitation.

Just over 70% of the surface of Earth is covered by water, but less than 3% of Earth's water is fresh, and three quarters of that is frozen at the poles. Of the remainder most is underground. Even though Earth is a watery planet less than one hundredth of one per cent is available for all life on land. Of this the majority is used for crop irrigation and industry with only 5% of world life consumption being used for human domestic purposes. Earth is very much divided into two as far as water availability is concerned.

Since 1950 the use of water in the world has increased by over three and a half times. At present two billion people (about 40% of the population) in 80 different countries are chronically short of water. The situation is getting worse. For example, by the end of the 1990s Kenya may only have half as much water for each of its inhabitants as it had in 1990.

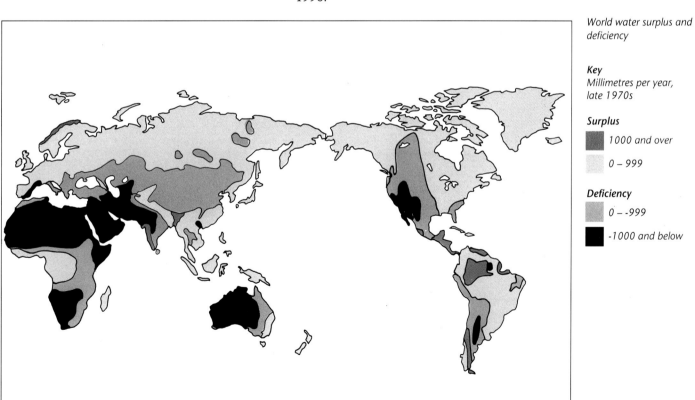

World water surplus and deficiency

Key
Millimetres per year, late 1970s

Surplus
1000 and over
0 – 999

Deficiency
0 – -999
-1000 and below

EARTH : **GLOBAL DECISIONS** 15

The following article is a summary of an article highlighting some of the water problems in Southern Africa.

The climate of Southern Africa is naturally dry but the region is suffering from a severe drought. The area's growing population is desperately short of water. The Zambesi river is being looked at as a potential solution to the problem in many of the neighbouring countries.

The Zambesi river

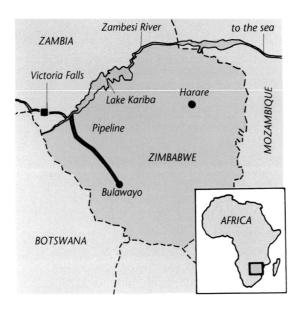

Five main reservoirs supply the city of Bulawayo with water. Drought has meant that these sources have dried up. As an emergency measure, to provide water for the city's population of one million, 450 boreholes have been drilled. Zimbabwe has long term plans to pipe water to Bulawayo from the Zambesi River. If this pipeline can be built Zimbabwe would become self sufficient in staple diet foods, with the necessary water available to be used for irrigation in agriculture. The lack of water in the region has caused the level of Lake Kariba to fall. This has in the past supplied both Zambia + Zimbabwe with electricity. Taking water from above the lake would seriously affect this operation. Zambia has planned to dam the river higher up to ensure hydroelectricity for the country.

South Africa is 1000 km from the river, but is another country needing water. It has plans to pipe water from an area upstream of the Victoria falls through Botswana. Implementation of this plan would necessarily affect the plans of the countries downstream.

(Adapted from *New Scientist*, 22 August 1992)

B. Discuss

- Read the information about the water problems in South Africa and then discuss the following:

  1. What factors are contributing to Zimbabwe's water shortage?

  2. Where does Bulawayo usually get its water from?

  3. Why is the emergency measure to drill many boreholes unlikely to provide a long term solution to the problem?

  4. How do Zimbabwe's plans for extracting water from the Zambesi affect Zambia?

  5. The border of South Africa is 1000 kilometres from the Zambesi River. The country has plans to pipe water from upstream of the Victoria Falls. What are the implications of this for the neighbouring countries?

It is estimated that at least 25 000 people die from using dirty water each day. In global terms there are too few taps. The World Health Organisation has stated that:

'*The number of water taps per 1000 persons is a better indication of health than the number of hospital beds. It is not uncommon in some parts of Africa for women to walk for seven miles to collect water. Yet it has been estimated that to provide everyone with clean water and sanitation would cost less than 10% of what the developed world spends annually on alcohol.*'

Section 2.2

MANAGING THE DROUGHT

Great Britain is in a humid region of the world. The average annual rainfall is about 750 mm. Precipitation is so common that we take it for granted but many areas of the country have experienced comparative drought conditions recently. This has led to an increased awareness of the need for management and development of the hydrological cycle.

In 1992 Great Britain entered the fourth year of the longest drought this century. During the last fifteen years droughts have also occurred in 1975, 1976 and 1984. The 1992 drought is unique in living memory as it was the longest. Not all areas of the country suffered in the same way. It was most severe over those parts of England with high areas of population, intensive agriculture and industry. Rainfall over the whole of Great Britain during this time was very close to the long term average. North West Scotland received 120 per cent of their average rainfall and had to cope with several severe floods while parts of the South and East of England received less than 80% of their average rainfall.

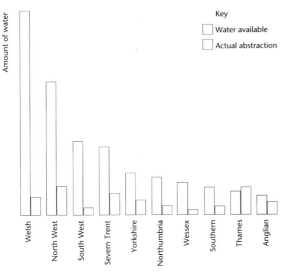

A comparison of the amount of water available for extraction and actual abstraction at present in the regions covered by the Water Service Companies (no account is taken for the re-use of water)

The water shortage problem is severe in the region covered by Anglian Water. This is naturally Britain's largest, flattest, driest region. It also has a rapidly increasing population.

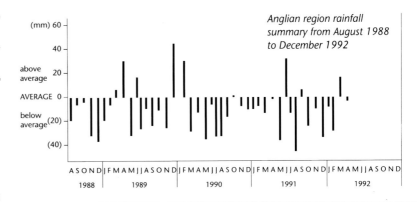

Anglian region rainfall summary from August 1988 to December 1992

C. Work out

- During the period shown on the block graph:

1 For how many months was rainfall:

   a below average; b above average?

2 What is the 'cumulative deficit' for this period?

3 The average rainfall of East Anglia is 600 mm. How many months rainfall is this 'cumulative deficit' equivalent to?

4 When considering water resources, planners use the term 'effective rainfall'. This is rainfall minus evaporation.

Copy and complete this table to compare the effective rainfall in East Anglia with that in the rest of England and Wales:

	Anglian region		Rest of England and Wales	
	Average year	Driest year (1 in 50)	Average year	Driest year (1 in 50)
Rainfall (mm)	595	463	940	770
Evaporation (mm)	448	423	453	450
Effective rainfall (mm)	147			

(1 in 50 is the extreme drought event that water schemes must be designed for)

5 Summarise the main reasons why East Anglia is running short of water.

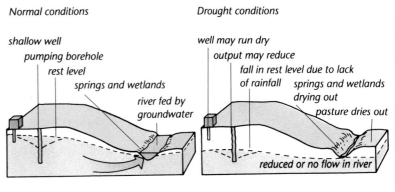

East Anglia gets much of its water from underground reserves

Average water consumption of household appliances

washing machine	100 litres
dishwashers	18-15 litres
flushing a toilet	8-10 litres
shower	30-63 litres
bath	80 litres

D. Work out

- Carry out a water audit either of a typical day at home or a typical day for your school. It would help if you entered this data into a spreadsheet.

- Think carefully of all the ways in which you could reduce the amount of water that you use. For example, do you wash up in a sink? If so, how much less water would you use if a bowl was used instead? Could this water then be used for something else, such as watering plants?

  Flushing a toilet uses a lot of water, how much water would be saved if a brick or two was added to the cistern? What is the biggest article that could be added to the cistern and the toilet still flush effectively?

- List as many ways as possible to save water. The best ways will be simple to operate while still maintaining the current lifestyle.

- Try out these ideas and then carry out a further water audit to see how effective you have been.

- Design a simple leaflet that could be sent to each household. The leaflet should explain why there is a need to conserve water and give simple instructions for ways of doing this.

One of the options for providing East Anglia with more water

Planning for the future

It is the duty of the National Rivers Authority *to conserve, redistribute or otherwise augment water resources ... and to secure their proper use.*

There are a number of options for helping to solve the particular problems of the Anglian Region.

1 Inter-regional transfer

East Anglia is the driest area of the country. One of the options is to move water from some of the wetter parts of the country.

This scheme would involve the building of a large new dam at an existing reservoir in Wales if Mid Cambrian Sources were used. The water would need to be transferred from this region by pipes or the existing canal network. It would also involve the building of storage systems in the regions receiving the water. The Kielder reservoir was completed in 1982. It was built to meet the predicted demands of the north-east region but the demands have been less than predicted so the reservoir has spare capacity. This water could be transferred by river to the Anglian region.

EARTH : **GLOBAL DECISIONS**

2 Re-use of effluents

The Thames region already re-uses effluent with approximately 13% of water already having been used before. This is a possibility for other regions. If water is re-used great care must be taken to protect public health.

3 Desalination

Aerial view of a desalination plant in Nevada, USA

There are a number of disadvantages in obtaining drinking water from sea water:

- fossil fuels are burnt to produce the water;
- the capital and operating costs are high;
- the water produced has a low concentration of salts, is unpleasant to taste and it needs mixing with water from more conventional sources;
- disposing of the concentrated brine from the plant poses problems.

At present the Channel Islands have the only public water supply desalination plant in Great Britain. It is not considered to be a practical solution for other places.

4 National Water Grid

The idea of a National Water Grid, which would operate in the same way as the National Electricity Grid, was first suggested in the 1940s. This would mean laying many miles of pipes, and building pumping stations to redistribute the countries water. The cost of this option is so high that it is not a viable proposition at the moment.

5 Water from Europe

The region on the European side of the Channel Tunnel has a surplus of water. It has been suggested that this could be piped through the tunnel to the regions of Great Britain that require it. The two main disadvantages of this proposal are:

- the tunnel was not constructed with this in mind – there is not sufficient space available;
- this region of France is an economic development zone and it is likely that industries which set up here may need the water in the future.

6 Transfer by ship

This option includes the movement of fresh water and the towing of Arctic icebergs. Both options would require long pipes to be laid from the docks on the East Coast. The shipment of fresh water has fewer disadvantages than obtaining water from icebergs but neither of these options are likely to be developed for long-term water supplies.

7 Using storage drogues

These are large storage vessels for water at sea. The idea is that these are filled with rain water in times of plenty, for example in the winter, and then this water could be used when needed. Although a possible option, this is probably uneconomic.

E. Discuss

- Consider each of the options available for the long term supply of water to the Anglian region. Make sure that you list the advantages and disadvantages of each option.

- Prepare a ten minute insert for the local news in the Anglian region to be broadcast during a particular dry spell. You should include:

  the fact that the Anglian region has scarce water resources;

  that the demands for water are rising;

  the potential alternative water resources (and cost!).

Section 2.3

WATER AS A SOLVENT

You will probably have used water as a solvent already today. Every time you make a cup of tea or coffee, a solute in the tea leaves or coffee grains dissolves in water (the solvent) to form a solution. Water is called 'the universal solvent'. Most biochemical reactions in living organisms occur in water and it is widely used in industry as a solvent.

F. Plan and test

- Compare the solubility of the given substances in water and 1,1,1-trichloromethane.
  You will need to decide how you will record solubility and whether there are any energy changes.

- When you have obtained your results, look for any patterns in the data and try to explain these using your knowledge of the substances.

If you build a model of water using two hydrogen atoms and one oxygen atom you will notice that the three atoms are not in a straight line but at an angle. The atoms in the water molecule are held together by covalent bonds. Four of the outer electrons of oxygen are not used for bonding. These four electrons are arranged as two pairs, called lone pairs.

Molecular model of water

a Two pairs (4) of the outer electrons of oxygen are used to form bonds with the two hydrogen atoms. The other two pairs of electrons are called 'lone pairs'

H : O :
　　‥
H

b This part of the molecule is 'more negative' than the rest because of the lone pairs

c Two ways of representing the uneven distribution of charge in a water molecule (∂^- means a very small amount of negative charge and ∂^+ means a very small amount of positive charge)

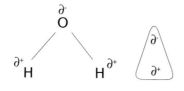

These lone pairs occupy the space in the model. Part of the molecule is more negative than the rest.

Water seems to dissolve virtually anything to some extent! It is particularly good at dissolving ionic substances.

The following diagram shows how polar water molecules attach themselves to ions in a solid crystal lattice. The ions then move away and the crystal dissolves.

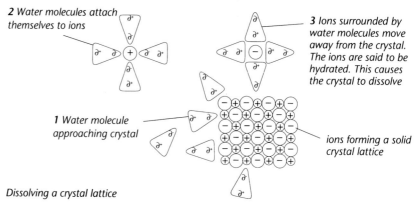

Dissolving a crystal lattice

Bonds are being broken and then made in this process.

1. Breaking bonds. This happens when the ions in the crystal lattice separate into free ions. Energy is absorbed as the bonds are broken.

2. Making bonds. This happens when new bonds are made between the ions and the water molecules. This is called hydration and results in the release of energy.

Depending on which of these two processes predominates, the overall reaction is endothermic or exothermic.

The following diagram shows the energy changes that occur when sodium chloride is dissolved in water.

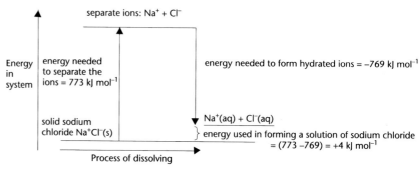

The process of dissolving sodium chloride is endothermic. More energy is needed to separate the ions than is released when new bonds are formed with the water molecules

Water is also able to dissolve some substances that do not have ionic bonding, for example glucose. Glucose contains oxygen and hydrogen atoms which form hydrogen bonds with the water molecules.

glucose molecule

Hydrogen bonding

Water for dyeing

One of the reasons why water is so useful is because it is a good solvent. This useful property, however, is not without its disadvantages. If you look around you now you will see many coloured items. Most of these colours are due to dyes. To be a good dye a coloured substance must attach itself to the substance that needs to be coloured, for example paper, hair, clothes and food. Dyeing uses a lot of water and the quality of the water is very important. The pH will affect the colours resulting from the dye, a neutral solution being the most versatile. If water dissolves a lot of ions as it travels through the ground it becomes what is called 'hard water'. Hard water with a high ion content can cause an uneven distribution of dye.

The dyers' market in Marrakesh, Morocco

G. Work out

1. Build a molecule of water and use the information from the diagram to explain why water is described as a polar molecule.

2. Fill the burette with water. Arrange for a very fine stream of water to run from it. Now take a plastic biro and 'charge it' by rubbing it on a jumper sleeve. Immediately place the 'charged' biro near to the stream of water. Note what happens.

   Repeat this using 1,1,1-trichloromethane in the other burette.

    You should do this in a fume cupboard.

   Explain your observations using your knowledge of the structure of water.

3. Explain why water can dissolve:

   a ionic solids;

   b some covalent solids.

4. a Draw a sketch to show the energy changes involved in dissolving each of the following:

      lithium iodide:
      (heat of solution -47 kJ mol^{-1})

      potassium iodide:
      (heat of solution + 39 kJ mol^{-1})

   b Explain each of your diagrams in terms of bonding.

H. Investigate

- Jane and Steven Moss are knitwear designers who are considering dyeing their own yarn so that they can produce the range of colours they require for their fair-isle designs. They intend to set up a small factory and are looking at three possible locations, but the final decision depends on the quality of the water.

  Location one: a small country farm with lots of suitable outbuildings and a barn. The water can be drawn from a well. The underlying rock is granite.

  Location two: a large warehouse with large roof tanks, in a heavily industrial area.

  Location three: a mountainside cottage in a limestone area, with the potential of a stream supplying the necessary water.

  Imagine that you have been called in to investigate the three types of water and produce a report for the Mosses. You will need to summarise your findings briefly and then include the data on which you have based your conclusions as an appendix.

- To find out how hard water is, you can carry out a soap titration:

  put 25 cm^3 of the sample into a conical flask;

  fill the burette with soap solution;

  take the burette reading;

  run soap solution into the sample gradually;

  after each addition shake the flask;

  stop when you have a permanent lather on the surface of the sample;

  note the volume of soap solution used.

UNIT 3
THE WEATHER

Section 3.1

THE GLOBAL ENERGY BALANCE

The Earth rising above the moon's horizon as seen from an orbiting space craft. The moon is almost the same distance from the Sun as we are on the Earth

The surface temperature of the moon is, on average, -18°C, varying between 100°C in the Sun to -150°C at night. The temperature of the air just above the surface of the Earth is, on average, 15°C. Both the moon and the Earth receive their heat from the Sun, but the Earth is about 33°C warmer than the moon. Why the difference? It is because the atmosphere around the Earth acts as a blanket keeping it warm.

The temperature of the surface of the Sun is about 6000°C. The Sun is about 150 000 000 km from the Earth and so it is not surprising that only 1/2000 millionth of the Sun's radiation reaches the outer atmosphere of the Earth. Most of this is in the form of short wavelength radiation. 30% of the solar radiation travelling to the Earth is immediately reflected back into space by the atmosphere, and 20% is absorbed by the atmosphere. This means that only about 50% passes through the atmosphere and 20% is used to heat the Earth's surface. This energy is eventually returned, as long wave radiation – 14% to the atmosphere and 6% to outer space. About half of the radiation actually reaching the Earth (24% of the total) is used to drive the water cycle. Ultimately, all this energy is returned to the atmosphere as water condenses in clouds and gives out heat. Of the 50% of the energy passing through the atmosphere, the remaining 6% is used to heat the atmosphere just above the Earth's surface. Eventually this radiation is also absorbed by the atmosphere.

A. Work out

- Use the information in this section to complete Worksheet GD7 *Global energy transfer* to show what happens to the solar radiation reaching the atmosphere.

The global energy transfer cycle is very much a generalisation. Not every part of the Earth fits this energy balance exactly. Major factors affecting this are:

1 Distance of the Sun from the Earth: the shape of the Earth's orbit (an ellipse) means that the intensity of the radiation from the Sun varies during the year. When furthest away from the Sun it receives 7% less solar radiation than when it is closest. The difference in distance varies from 151 million kilometres to 147 million kilometres.

2 Altitude of the Sun: the angle between the Sun's rays and the Earth affects how much solar radiation hits the Earth.

X, Y and Z are bundles of rays with equal heating power when they reach the atmosphere

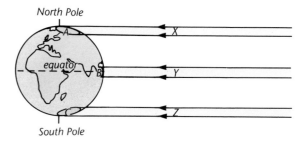

The heating effect of the Sun's rays depends on the angle at which they strike the ground

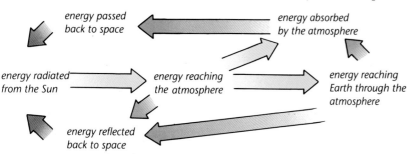

A simplified diagram of the global energy cycle

22 EARTH : **GLOBAL DECISIONS**

3. Length of the day: radiation from the Sun only affects the Earth during the day. The longer the day the more solar radiation hits it. This depends on the time of the year and the latitude of the country. In polar regions days are very long or very short. In equator regions they are always about 12 hours long.

Over 2000 years ago Aristotle thought that it was the difference in temperature between the poles and the equator which provided the necessary force for the general circulation of the atmosphere. In general his idea is still accepted today.

The equator region gains more energy than the polar regions. This makes the air circulate in the atmosphere to carry energy from the hotter to the cooler regions. A balance is set up so that any place on the Earth has a similar temperature from year to year.

B. Plan and make

- Plan and make a model or a poster that will show how the solar radiation to the Earth is affected by the factors listed.
- Include brief notes to explain each of the factors involved.

The greenhouse effect

As well as receiving heat, the surface of the Earth radiates heat. This is absorbed by the atmosphere and some is radiated back to the Earth to keep it warmer than it would otherwise be. This is the basis of the 'greenhouse effect'. Water vapour and carbon dioxide in the atmosphere absorb this radiation. As long as the amount of water vapour and carbon dioxide stay the same then there is a radiation balance and the temperature of the atmosphere does not change much.

The 'greenhouse effect' has caused concern since the 1860s when British, American and Swedish scientists were worried about the effects of water vapour and the increasing amount of carbon dioxide being produced by the burning of coal as a result of the industrial revolution. Since that time the mean temperature has increased but there is no proof that this is just because of human activity. Measurements are now usually taken at the South Pole. These are as accurate as possible because the atmosphere there is not close to pollution and has had a chance to get well mixed up.

In the middle of the nineteenth century the natural concentration of carbon dioxide was taken to be 270 ppm. Later measurements of the amount of carbon dioxide trapped in the polar ice sheets confirmed that this concentration had persisted throughout the past 10 000 years. This is taken as the baseline against which measurements and predictions are made.

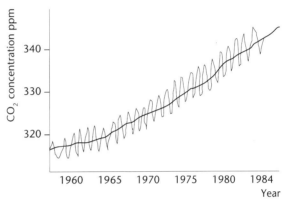

Although there is a seasonal variation in the amount of carbon dioxide in the atmosphere, the average amount does show an increase
(The data for this graph was recorded at the Mauna Loa Observatory, Hawaii)

Many different predictions of the effect of this increase of carbon dioxide in the atmosphere have been made using computer models. The models all differ slightly but the predictions have a number of common features:

1. Doubling the amount of carbon dioxide from 270 ppm will increase the mean temperature of the Earth by about 2°C;
2. Warming will be greatest at the Poles;
3. The direction of the prevailing winds will change;
4. The areas that receive most and least precipitation will change.

If the present situation continues, it is estimated that the carbon dioxide in the atmosphere will double by the year 2080. However, other gases contribute to the greenhouse effect so the changes could come much earlier.

More coal is being burnt each year. For each tonne of coal burnt almost four tonnes of carbon dioxide are produced.

$$C + O_2 \longrightarrow CO_2$$

However, all the carbon dioxide produced is not just added to the atmosphere. It is thought that up to half is absorbed naturally, by plants on land and plankton in the oceans, during normal photosynthesis. Large amounts of carbon dioxide are also probably dissolved in the oceans which cover 80% of the Earth.

The following table gives some information on the other greenhouse gases.

Gas	Source	Each molecule compared to carbon dioxide	Greenhouse effect total contribution in the 1980s
CFCs	• aerosols • refrigerants	15 000 times greater	14%
methane	• bacterial activity in paddy fields and cows • release of natural gas • from oil and gas fields	30 times greater	18%
dinitrogen oxide	• burning vegetation • given off by fertilisers	150 times greater	6%
ozone	• sunlight on gases produced from vehicles	2000 times greater	12%

The actual amounts of these gases are very small compared to carbon dioxide: their *total* contribution in the 1980s was equal to that of carbon dioxide.

By calculating the percentage increase of these gases and their greenhouse effect, it is predicted that the 2°C increase in temperature may well occur 50 years sooner than that predicted by considering carbon dioxide only.

C. Present

- For your school's open day, the science department is going to put on a number of exhibitions, talks etc. about items of interest involving science. Your group has been asked to give a presentation with the title: 'Greenhouse effect; fact or fiction'. This could be in the form of a set of posters, tape-slide sequence, ten minute talk, leaflet etc.

  You will need to use the information in this section and, if possible, to research the subject more thoroughly. It is important that you include known scientific facts and present them clearly. Remember some of your audience may not have done any science since they were at school.

Section 3.2

AIR IN MOTION

Our weather is due to energy from the Sun reaching the Earth. More of this energy is absorbed at the tropics than at the poles. In very simple terms, the atmosphere moves to try and smooth out these differences in energy distribution. There are patterns in how the atmosphere circulates, and understanding these enables us to forecast the weather with some accuracy.

In 1863, Admiral Fitzroy suggested that the weather of Great Britain depended on what happened when 'airstreams' met. Today we call these 'air masses' but meteorologists agree that they have an important part to play in helping to understand the weather. At that time most meteorologists concentrated on describing the air masses. The three main source regions of these are known as arctic (A), polar (P) and tropical (T) and these are then further subdivided into continental or maritime, depending on whether they have come from the land or the sea. The arctic air mass is not usually subdivided as such so, in fact, this results in five air masses. Each mass is relatively uniform and can be kilometres across.

It has been found that air masses move away from the 'source area' where they were created

Air mass	Source region	Temperature	Humidity
arctic (A)	within the Arctic Circle	very cold or severe in winter; cold in summer	dry
polar maritime (Pm)	ocean in vicinity of Iceland and Greenland polewards of 50°N	cold in winter, cool or rather cold in summer	rather moist
polar continental (Pc)	Northern Eurasia	cold or very cold in winter; warm or very warm in summer	dry
tropical maritime (Tm)	Azores	mild in winter; warm in summer	moist at surface; dry high up
tropical continental (Tc)	North Africa and Mediterranean	mild in winter; hot in summer	dry

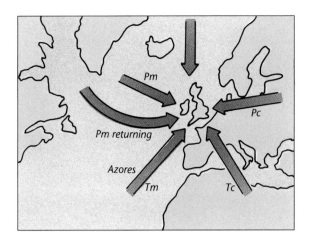

The five air masses which influence our weather

polar air that is travelling from the poles. The boundary between this cold polar air and the warmer air moving above it is known as a **polar front**.

Great Britain sits under the polar front region which is the boundary between the cold polar air mass and the tropical air mass. It is this zone which has such an influence on the weather of Great Britain. It is here where the frontal depressions form.

but tend to maintain their characteristics. For example, the air masses from the arctic stays cool and the tropical air masses stay warm. When air masses of different characteristics meet they do not mix easily, but tend to keep separate with a sloping boundary between them. This boundary is called a front. These fronts are of great interest in weather forecasting as it is here that changes in the weather occur. Research into the fronts forming in the mid-latitude position (which ultimately result in our weather) has resulted in the polar front theory. This theory has helped to make weather forecasting in Great Britain more accurate. The following describes the main ideas in the theory:

1 The atmosphere would not move at all and there would be no winds if:

the surface of the Earth was composed of uniform material (e.g. all land of the same height);

the Sun heated the Earth equally;

the atmosphere was of the same depth over the Earth.

2 The Sun heats the equator more than the poles of the Earth, so it can be predicted that the air must circulate in the atmosphere. At the equator the warm air rises and at the poles the cool air sinks. The pressure at the equator is low so the air rises and air from the poles moves in to replace this moving air.

3 As the hot air from the equator travels to the poles, it falls when it reaches the mid-latitude position. As it approaches the surface of the Earth it spreads. Some returns to the equator and some rises above the

Polar fronts form when masses of cold and warm air meet out in the Atlantic, usually at a latitude between 45° and 70°

4 The rotation of the Earth affects how the atmosphere moves.

Convection currents and the rotation of the Earth results in a more complicated circulation of the atmosphere

a *If the Earth did not spin, then the tropical air would rise and then flow due north and due south*

b *Because of the spin of the Earth, the air flow is deflected eastwards*

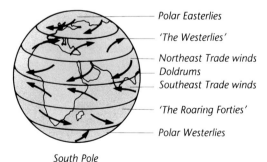

c *Mountains and sea cause other convection currents which affect the global movement of air and make the patterns more complex*

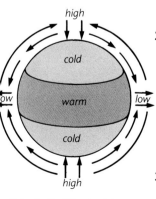

If the Earth did not rotate and was made of uniform material the atmosphere would be like this

If you study weather forecasts you will soon see that in Great Britain most of our weather systems come from the west. This is because the prevailing winds are from the west – but why? Each point on the surface of the Earth rotates eastwards once in 24 hours. A point at the equator moves at 470 metres per second. At a latitude of 60° (by Great Britain) a point only rotates at 224 metres per second, and nearer the poles a point rotates even slower. This means that air rising from the equator northwards to the North Pole is travelling east faster than the land. It therefore produces winds in the mid latitudes which come from the south west. Air returning to the equator southwards has to travel over land that is moving faster than it. This means it produces winds from the north east.

The weather in Great Britain is brought predominantly from the west, and is the result of the passage of depressions over the country. Looking at weather maps on consecutive days provides evidence for this.

D. Work out

- Assemble the apparatus as shown in the diagram. Light the candle and hold a smoking taper above the chimney, as shown.
- Draw a diagram of the model. Annotate it to show how it can be used to explain simply the movement of the atmosphere. You will need to think about which parts represent:

  the equator;

  one of the poles.

- Which part of the model would be a mid-latitude region?
- Where would a polar front form?

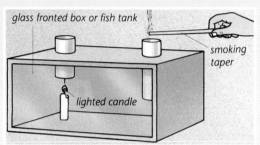

- Suggest modifications to the model that would make it more realistic.

Depressions are areas of low pressure which develop along the polar front. This occurs when buckles begin to form along the straight polar front between the rival air masses, with the warm air displacing the cold air in some places and the cold air displacing the warm in others. Sometimes the air pressure begins to fall rapidly at the neutral point between the conflicting air masses, known as the wave tip.

Monday 18 January 1993

Tuesday 19 January 1993

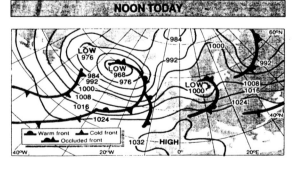

Section through the weather system (side view)	As shown on the weather map (plan view)	Description
1 	cold front / warm front	Warm air moving north bulges into the cold air moving south, forming a warm front to the east and a cold front to the west.
2 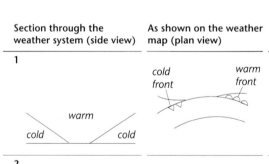	low	The bulge gets bigger and the warm air rises over the cold air because the warm air is less dense. As it rises it cools and clouds form. Localised winds start circulating around the wave tip in an anti-clockwise direction.
3 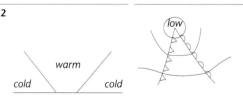	occluded front / low 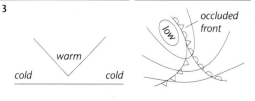	The cold front moves faster over the Earth's surface than the warm front. In time the cold front catches up with the warm front, and squeezes the warm air upwards until it is lifted away from the surface altogether. This is known as an occlusion.

Stages in the development of a depression

Comparing warm and cold fronts

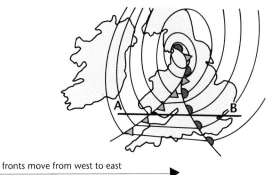

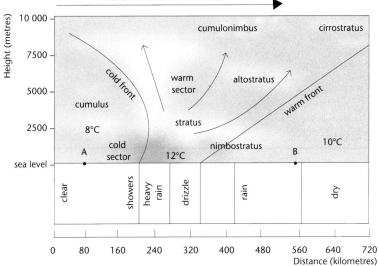

Element	Before	At front	After
Warm front			
pressure	steady fall	fall ceases	little change or slow fall
wind	increases and sometimes backs	veers and sometimes decreases	steady direction
temperature	steady/slow rise	rise	little change
cloud	Ci, Cs, As, Ns in order	low Ns	St
weather	continuous rain (or snow)	precipitation almost stops	cloudy, drizzle or light rain
Cold front			
pressure	fall	sudden rise	rise continues more slowly
wind	increasing and backing, becoming squally	sudden veer, perhaps squally	backing a little after squall, then steady or veering
temperature	steady, but falls in rain	sudden fall	little change, variable in showers
cloud	Ac or As, then Cb	Cb	lifting rapidly, followed by As or Ac; later further Cu
weather	rain	rain, often heavy, with perhaps thunder and hail	heavy rain for short period, sometimes more persistent, then fair with occasional showers

Key
Ci = cirrus
Cs = cirrostratus
As = altostratus
Ns = nimbostratus
St = stratus
Ac = altocumulus
Cb = cumulonimbus
Cu = cumulus

When this happens localised winds start circulating around this point – always in an anti-clockwise direction in the northern hemisphere – and a low-pressure area, or depression, is formed.

Since the use of satellites in the early 1960s it has been possible to see cloud patterns and to find out more about depressions formed along the polar front.

When an occlusion forms the cloud mass develops a characteristic comma shape.

A receding occlusion

E. Work out

- Complete Worksheet GD8 *Interpreting depressions* to summarise the weather associated with a depression.

In the centre of a depression, the air flows towards the centre, or converges. This air then rises into the upper regions and diverges (or spreads out). Most frontal depressions (also called cyclones) nearly always form on the easterly side of a trough in the upper westerlies.

Areas of high pressure are most likely to occur in the western side of such a trough. These are associated with anticyclones. The term was first used in 1861 to describe weather associated with concentric circles around a high pressure area. These systems tend to have the opposite characteristics to depressions and so are called 'anticyclones'. Anticyclones are usually larger than depressions, up to 3000 km across, slower moving and more persistent than depressions. They are associated with relatively clear skies and dry settled weather.

EARTH : **GLOBAL DECISIONS**

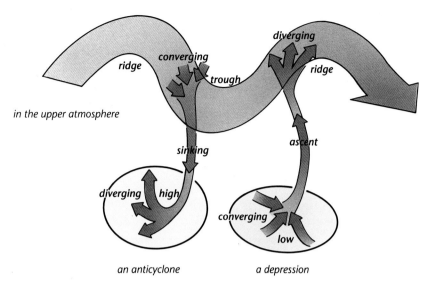

The formation of an anticyclone and a depression

Feature	Anticyclone	Depression
pressure at Earth's surface	high	low
wind direction	anticyclonic (clockwise*)	Cyclonic (anticlockwise*)
airflow	diverges at surface (converges high up)	converges at surface (diverges high up)
vertical air motion	subsides	rises
wind speed	weak	moderate to strong
precipitation	generally dry	wet
cloudiness	stratus or no cloud	cloudy
temperature gradient	little temperature contrast across the high	strong temperature contrasts, especially at the fronts
speed of movement	slow-moving or stagnant	generally mobile, moving west-east

* In the Northern hemisphere

Comparative features of an anticyclone and a depression

F. Work out

- Copy this table and complete it to show the differences between depressions and anticyclones.

Property	Depression	Anticyclone
pressure at the Earth's surface		
direction of wind		
vertical air flow		
amount of cloud		
precipitation		
speed of movement		

All the time researchers are aiming to understand more about the causes of the weather. This research is based on:

observational studies, actually watching how the atmosphere circulates in the upper and lower atmospheres;

laboratory models, the so called 'dishpan' experiments in which columns or circular dishes of water, simulating the atmosphere, are rotated and heated to simulate the activity of the atmosphere at the equator and the poles;

computer modelling in which lots of data can be fed into mathematical equations of the known laws governing atmospheric motion.

The global circulation of the atmosphere is very complex, but this research supports the idea that the behaviour of the air at the polar front, causing the familiar depressions of our weather forecasts, is very important in the transfer of energy over Great Britain.

A climatic physicist using portable air sampling equipment

Section 3.3

WEATHER FORECASTING

Have you noticed the weather today? Can you remember what it was like when you woke up?

Great Britain has a temperate climate, so in general we do not have the extremes of weather experienced in some countries.

You may only think about the weather when you are planning what to do in your spare time. Weather used to have a bigger effect on the everyday life of people than it does today. The following is an extract from the Reverend F Kilvert's diary. He describes Christmas day in 1870:

Sunday, Christmas Day

As I lay awake praying in the early morning I thought I heard a sound of distant bells. It was an intense frost. I sat down in my bath upon a sheet of thick ice which broke in the middle into large pieces whilst sharp points and jagged edges stuck all round the sides of the tub like chevaux de frise, not particularly comforting to the naked thighs and loins, for the keen ice cut like broken glass. The ice water stung and scorched like fire. I had to collect the floating pieces of ice and pile them on a chair before I could use the sponge, and then I had to thaw the sponge in my hands for it was a mass of ice. The morning was most brilliant. Walked to the Sunday School with Gibbins and the road sparkled with millions of rainbows, the seven colours gleaming in every glittering point of hoar frost. The Church was very cold in spite of two roaring fires. Mr. V preached and went to Bettws.

(from *Kilvert's Diary*)

The weather may not influence you as much as it did the Reverend Kilvert! However some industries, such as agriculture and transport, are very dependent on the weather. The food industry is also interested in the weather and predicted weather – ice cream companies need accurate forecasts to plan production and distribution.

G. Discuss

1. How has the weather affected you today?
2. When has the weather affected you most?
3. How would the weather have affected you if you had lived in the sixteenth century.
4. What industries (other than agriculture, food and transport) are strongly influenced by the weather?

The sale of ice-cream often depends on the weather

Weather forecasting depends on interpreting past weather and using this information to predict future weather.

Theophrastus, a Greek Philosopher may have been the first weather forecaster. He noticed patterns in the weather, for example that some winds were more likely to bring rain than others. He tried to explain this and realised that these winds were either coming from the sea or from over mountains. Farmers and sailors in particular continued to have their own methods of forecasting the weather. In the time of Elizabeth I, the admiralty kept a 'wind book' but there was no recording of the weather, in quantitative terms until the invention of the thermometer and barometer.

In 1643 an Italian, Torricelli, invented the barometer. With this he showed how atmospheric pressure changed.

Robert Hooke realised that to produce weather forecasts with accuracy, measurements must be taken of the atmospheric conditions at any given time. In about 1670 he suggested that the weather was related to the pressure of the atmosphere.

He thought that low pressure meant stormy weather and rain. Even today barometers are made which show this, although we now know the connection is not that strong.

At about the same time, the first 'liquid-in-glass' thermometer was invented, probably by Galileo. This was also used by Robert Hooke in making his weather records.

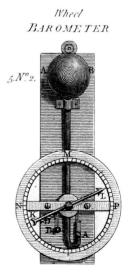

Robert Hooke's wheel barometer

In more recent times interested individuals measured rainfall and humidity (the amount of water in the air), wind direction and cloud type and cover. For many year there were recordings only of separate places.

A breakthrough in weather forecasting came with the ability to collate and analyse data collected in separate parts of the country.

The first weather map was not produced until the 1840s. It was then that the telegraph was invented and information collected over different parts of the country could be processed fairly quickly.

Weather map of 1851

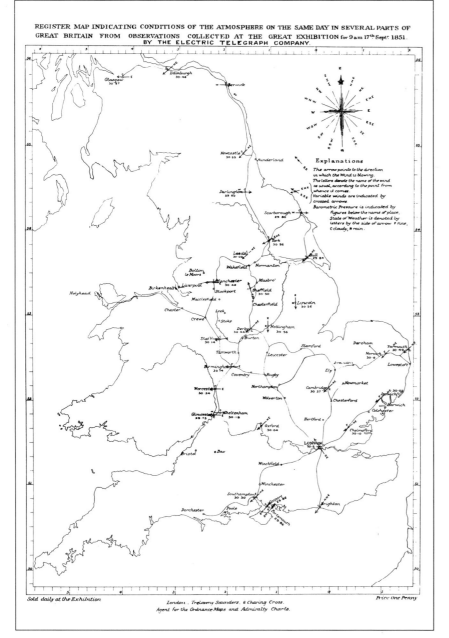

From 1823 groups of people met as 'Meteorological Societies', but it was not until 1854 that the Meteorological Office was set up. It was in the 1870s that daily weather forecasts as we know them today were issued.

H. Design

- Imagine that a company wishes to develop some simple weather recording equipment to be used by Junior Schools. The pupils will be using the equipment to help in weather forecasting. They are interested in apparatus to measure:

  rainfall;

  temperature;

  wind speed;

  wind direction.

  You have been consulted on this proposal, the company wishes to establish some facts before commencing. You are asked to advise on whether or not these are the most important instruments and, if not, should some be omitted and others included.

  Assuming that the company wishes to go ahead with the development, you are asked to advise on:

  1. the design of each of the instruments;
  2. suitable operating instructions (remember who will use the instruments);
  3. a suitable kit in which to package and use the instruments;
  4. helpful suggestions about recording and interpreting the data.

Nowadays, weather data collected on the ground is supplemented by data obtained from radio-sound devices which are sent out daily.

Weather satellites add another dimension to the accuracy of weather forecasting in that they can provide global information. About three quarters of the Earth's surface is covered by ocean, desert or polar regions. It is much easier to obtain information from these regions by satellites. This is very important as these regions are the source of much of the Earth's weather.

Satellites are used to record cloud patterns as well as ground temperatures and data on the conditions throughout the levels of the atmosphere.

Computer operating room at the Met Office

All this information is fed into computers which are able to run simulations which model the conditions and so the accuracy of weather forecasting is rapidly improving.

I. Research

- Collect as many different weather forecasts as you can for one particular day.
- Evaluate the forecasts and find out which is the most accurate.

  Is this source the most accurate for a whole week?

- Collect sufficient data to enable you to say which *is* the most accurate source of weather forecasting.

J. Interpret

1 Look at these weather maps from foreign newspapers:

Predict the weather at London for each weather map.

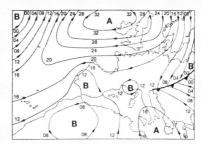
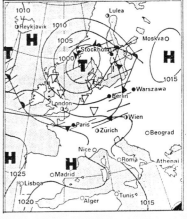
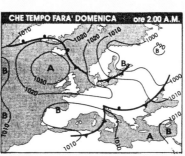
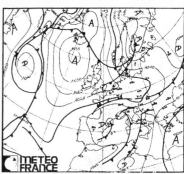

2 Look at this weather map:

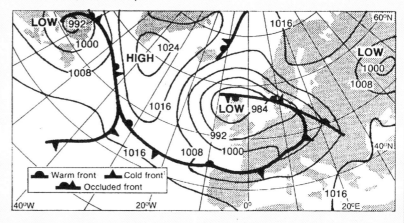

Describe where you would expect to find the following phenomena:

a cirrus clouds;

b barometers falling;

c very windy conditions;

d the temperature rising rapidly;

e a possibility of thunder and hail;

f the wind blowing from a northerly direction;

g a warm front;

h a cold front;

i an occluded front;

j air moving vertically upwards.

EARTH : **GLOBAL DECISIONS**

UNIT 4
UNDERSTANDING THE ENVIRONMENT

Section 4.1

ECOSYSTEMS

We can study the environment in terms of its plants and its animals, and their non-living surroundings.

Animals and plants interact with each other and with their non-living surroundings to form a natural unit called an ecosystem.

A. Discuss

- Look at the parts of the ecosystem in the illustration.
- Make a list of how each part of the ecosystem interacts with the other parts.

  For example, one part eats another part!

B. Research

- Choose another ecosystem and list the animals, plants and the non-living surroundings found there.
- Make a poster to show someone how the parts of the ecosystem interact with one another.

32 EARTH : **GLOBAL DECISIONS**

Section 4.2

ADAPTING TO ECOSYSTEMS

There are many types of ecosystem. There are land-based ones like woodlands, meadows, heaths, deserts, rain forests and savannahs. There are aquatic ones ones like ponds, streams, lakes and seas. There are also many which are a combination of land and water such as seashores, bogs and swamps.

However, most ecosystems are based on a similar plan. There are plants, plant-eating animals and predators. How ecosystems differ is in the number and types of these groups of organisms and also in the type of environment in which the organisms live.

The photograph shows some of the plants that are common in a British woodland ecosystem:

These are some of the plant-eating animals found in a woodland:

These are woodland predators:

Animals and plants are adapted to the role they play in nature. The plants' role is to absorb solar energy, and so plants have a large surface area to achieve this. Many plant-eating animals have specialised teeth or mouthparts to snip off and grind up tough plant material. Others have tube-like mechanisms to reach nectar in flowers or to suck sap from the plants. Plant-eating animals are the prey for predators and so they must have mechanisms to detect predators approaching them. They also need some way of making a quick escape.

Predators need a means of finding and catching their prey. They also have specialised teeth and gut to break up and digest animal flesh.

C. Discuss

- Study the plant-eating animals in the illustration and discuss the ways in which they are specialised.
- Study the predators in the illustration and discuss the ways in which they are specialised.
- Make a list of the plant-eating animals that could be found in the savannah of Africa.
- Discuss the ways in which the African plant-eating animals are similar to and different from the woodland plant-eating animals.
- The savannah of Africa has a dry climate. The vegetation is mainly tough grasses and some trees. Make a list of the predators that could be found there.
- Discuss the ways in which the African predators are similar to and different from the woodland plant-eating animals.

Animals and plants are also adapted to the conditions in which they live. Even the harshest conditions can support life. The plants in the photographs are adapted to conditions of low water availability. They are called xerophytes.

1 Cactus
2 Marram grass
3 Pine tree

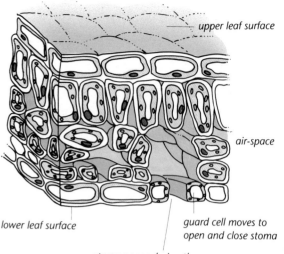

Section through a leaf showing the stomata

Water passes out of a leaf through tiny pores called stomata. The lower leaf surface usually has more stomata than the upper surface and so more water is lost from it. The number, size and pattern of stomata on a leaf varies from species to species.

Plants have a large surface area for absorbing solar energy, but this creates a problem for them as they also loose a considerable amount of water from this surface.

D. Observe and record

- Collect four leaves of a similar size from a tree or large plant.

- Cover the top surface of one leaf, the lower surface of another leaf and both surfaces of a third leaf with a thin layer of Vaseline. Leave the last leaf without Vaseline.

- Tie a length of string between two clamp stands. Tie all four leaves to the string and leave them for several days.

- Make a record of what happens to each leaf over the next few days.

- Explain what this experiment shows about water loss from leaf surfaces.

E. Observe and record

- Collect a number of different types of leaf.

- Paint acetone onto the upper surface of one of the leaves.

- Quickly press the leaf firmly onto a piece of acetate. This makes an imprint of the leaf on the acetate.

- Observe the imprint of the leaf with the x10 objective on the microscope.

- Make an imprint of the lower surface of the leaf and compare it with the imprint of the upper surface.

   How do the number, size and pattern of the stomata differ?

- Make similar imprints of other leaves and compare them with each other.

Leaves contain large air spaces. Water evaporates from leaf cells into the air spaces and then diffuses out through the stomata. The warmer the conditions, the faster the diffusion occurs. Plants living in hot places, such as deserts, would lose vast amounts of water by diffusion. However, plants that have evolved adaptations to reduce water loss can survive even in such harsh conditions. Cacti have greatly reduced the area of their leaves so that they have become simple spines. This reduces water loss. Cacti also use their fleshy stem to store water. Cacti stems also help the plant absorb light energy to compensate for its smaller leaf size.

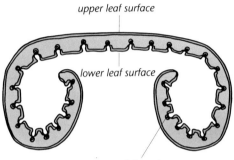

Vertical section through a marram grass leaf

Marram grass grows on sand dunes where water is in short supply. It is a xerophyte, which means that it is a plant specially adapted to withstand water shortage. It has an inrolled leaf with all of its stomata on the inside of the leaf. Its stomata are also sunk in pits on the leaf surface. Both the inrolled effect and the pits help reduce water loss.

F. Work out

- Study the illustration showing a section through a marram grass leaf.
- List the ways in which marram grass is adapted to reduce water loss from its leaves.
- Work out how the adaptations of marram grass reduce water loss from the leaf.
- Explain to someone how marram grass can grow successfully on sand dunes.

Section 4.3

POPULATIONS

Each ecosystem has its own type of plants and animals. It will also differ from other ecosystems in the number of organisms it can sustain. In harsh environments, such as polar deserts, only a few types of organism can survive, and these tend to be found in small numbers. In favourable environments, such as an established broad-leaved woodland, there are many different types of organism and these can be found in large numbers.

The type of environment determines the number of organisms that are found there

The number of a particular species in an ecosystem is called a population. While it might be easy to count the population of fish in an aquarium, the number of animals or plants in an ecosystem is usually difficult to count. This is either because there are a large number of most of the organisms, the organisms are small, or the animals can move around and hide. The only way to get an idea or estimate of the population size of each organism in an ecosystem is to sample. This means taking a small part of the ecosystem, counting the organisms in that area and then using that result to estimate the total population within the whole ecosystem. With animals you can use a technique called mark and recapture. With plants you need to use quadrats to look at the number of plants, or the area of cover, in an area within the ecosystem.

G. Investigate

- Use Worksheet GD10 *Mark and recapture* and/or Worksheet GD11 *Plant cover* to help you to estimate some populations in your local environment.

Population size is rarely static. New organisms are born or germinated while others in the population die. There are many factors in the environment that can influence population size such as the weather, environmental temperature, amount of food or energy available, number of predators, number of parasites or disease organisms and human influence.

It is now unusual to see British meadows with large numbers of meadow flowers, because of modern agricultural techniques

H. Observe and record

- Half fill a 250 cm³ beaker or similar sized jar with pond water.
- Place a twig in the beaker of water.
- Place one damselfly nymph in the beaker of water and leave it for about five minutes to settle down. Damselfly nymphs feed on water fleas.
- Using a pipette, transfer ten water fleas to the beaker containing the damselfly nymph.
- Leave the beaker for at least 20 minutes undisturbed.
- Meanwhile, discuss and make a list of the factors in a pond that could affect the number of water fleas and damselflies.
- Observe the beaker and count the number of water fleas.
- Pool the class results to get some idea on the predation rate of damselfly nymphs in the experimental conditions.

I. Investigate **EXTENSION**

- Plan and carry out an investigation to find out which factors greatly influence the population size of water fleas in a pond.

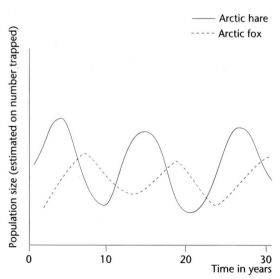

Variation in the population size of Arctic hares and Arctic foxes over a 30 year period

Food chains are often responsible for keeping population size under control. Predators keep the number of plant-eating animals down and they, in turn, graze on plants and reduce their growth and number. The interaction is far from simple, but it seems that in many cases each feeding level keeps a check on the next. This can be seen in the graph which shows how the population of Arctic hares and Arctic foxes varied over a 30 year period. Under favourable conditions the number of Arctic hares began to rise. There was a greater food supply for the Arctic foxes. After a time, this lead to an increase in the Arctic fox population. An increase in the Arctic fox population resulted in more Arctic hares being killed and eaten. So the population of Arctic hares began to fall. This cut the food supply for the Arctic foxes which eventually resulted in a decrease in the population size of the Arctic foxes. Over thirty years the populations of the two groups tend to show regular fluctuations that are slightly out of phase with each other.

The Great Barrier Reef in Queensland, Australia

Sometimes population size can change quite dramatically, and when this happens it usually greatly affects the ecosystem in which that population lives. An example of this has occurred in the last 50 years on the Great Barrier Reef, off the north-east coast of Australia. Coral reefs are formed by colonies of small animals called coral polyps. These form the food for starfish on the reef. The starfish tend to group together in batches of a hundred or so animals, feed voraciously on the coral polyps and then move to a new area of the reef. The starfish form the food of a large mollusc called the giant triton. Since the early 1960s, the population of giant triton has decreased greatly. This decrease is thought to be linked to increased pollution in areas neighbouring the reef. The pollution has led to a large increase in the number of starfish on the reef and severe damage to the reef.

J. Work out

1 Explain how the number of starfish on the Great Barrier reef were kept in check before 1960.

2 Explain why there has been damage to the reef over recent decades.

Once the balance of a population is lost it is often difficult for it to survive. This has happened with the African elephant. African elephants have been hunted for sport, for their meat and also for their ivory tusks. As more land and forest is taken for farming the elephants' territories have become much smaller and this has led to a reduction in the population size.

In some countries, large game reserves were set up with wardens to protect the elephants from poachers. Elephant numbers increased dramatically in the game reserves. However, this created further problems for the elephants. The game reserves were small in area and so the food supply for the elephants was limited. The elephants quickly ate all the available plants and then started to remove the bark and branches of trees in their search for food. Without plant cover, the land became desert. The elephants now face starvation.

K. Discuss

- List the stages that led to the African elephants facing starvation in game reserves.

- Discuss what actions the wardens should take to ensure that there are future generations of African elephants.

- Decide on the advantages and disadvantages of each of the actions that the wardens could take.

- Draw up an action plan that could save the elephants in the game reserves.

- Compare your action plan with those drawn up by other groups.

A hungry African elephant can destroy trees in its search for food

Section 4.4

ENERGY IN ECOSYSTEMS

A solar power station in California

All ecosystems on the Earth are powered by the Sun. In Section 3.1 you learnt how energy from the Sun interacts with the Earth's atmosphere. Half of the energy is absorbed or reflected back into space. The rest reaches the Earth in the form of heat and light energy.

Heat energy can warm up the non-living environment and also animals and plants. Some animals use this energy to increase their body temperature and so make it work more efficiently. Many lizards bask themselves on rocks during the day so that the heat from the Sun can warm them.

Green plants can absorb light energy from the Sun. They convert the light energy into stored chemical energy, which they use to fuel the many reactions that take place inside their cells. Plants are able to do this because they contain the green pigment, chlorophyll.

A lizard basking in the Sun

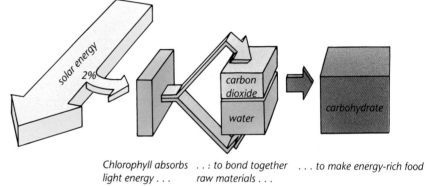

Chlorophyll absorbs light energy . . .   . . . to bond together raw materials . . .   . . . to make energy-rich food

In photosynthesis, energy is trapped and stored

In some countries, the heat energy from the Sun is used to heat water. This is achieved by using solar panels. Almost all the houses in Cyprus use solar panels to heat their water. Solar panels are also used extensively in Israel, Japan and Greece. Heat energy from the Sun can also be used in power stations to generate electricity. The first solar power station was completed in Israel in 1979. It used special 'ponds' which absorb and trap the energy in salt water. Since then, another 20 or so solar power stations have been built. Many of these are in the United States. They use vast U-shaped mirrors to direct and concentrate the Sun's rays onto pipes filled with water or oil. The energy transferred to the liquid in the pipes can later be utilised to drive electricity turbines.

Light energy is absorbed by chlorophyll to build simple raw materials into large energy-storing molecules. Carbon dioxide and water are built into sugars and starch, trapping and storing the energy. This process is called photosynthesis.

L. Plan and present

- Make a model or a poster to help you to explain to someone what happens to the Sun's energy that reaches the Earth.

Animals cannot photosynthesise and so rely on green plants for their energy. When a green plant is eaten, the energy-containing materials are transferred from the plant to the eater. Animals that eat plants are called herbivores.

38   EARTH : **GLOBAL DECISIONS**

Herbivores are eaten by other animals called carnivores. The energy-containing food in the herbivore is transferred to the carnivore. So, each time something is eaten, energy-containing foods are transferred from the food to the feeder. Energy is passed along a chain called a food chain.

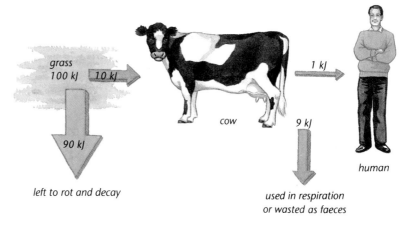

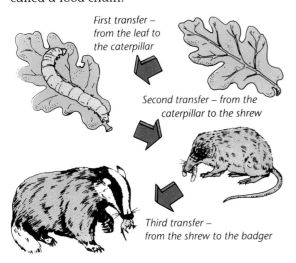

Only about 1% of the energy absorbed by plants reaches the human in this food chain

Energy transfer in a food chain

	Sheep	Cow
mass of grass consumed per day per animal	3 kg	70 kg
average total mass of animal	80 kg	600 kg
faeces per day per animal	1.2 kg	35 kg
urine per day per animal	1.3 l	20 l

As energy-containing materials are passed along a food chain, the amount transferred gets less and less. This is because some of the energy-containing materials are involved in driving the organism's life processes, some gets passed out as waste and the remainder is stored. Only the stored energy can pass along to the next organism in the food chain.

M. Work out

- Study the information given in the diagram about energy transfer through a simple food chain.

- Calculate how many kilojoules of energy are lost from the food chain when energy is transferred from:

  the grass to the cow;

  the cow to the person.

- Calculate the percentage energy loss at each transfer in the food chain.

- Make a list of the main ways in which energy is lost from the food chain.

- Using the data given in the table decide whether cows or sheep are better at converting plant material into animal flesh.

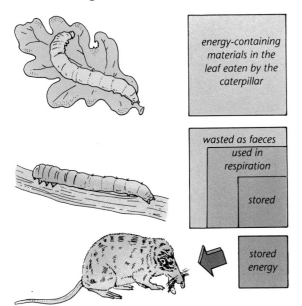

Only a small proportion of the energy-containing material is passed on to the next level in the food chain

Power stations can also be fuelled by plant material. Coal-fired power stations receive the energy that plants trapped thousands of years ago. However, in Hawaii, sugar residue is used to produce a third of the country's electricity, while India, Malaysia and Thailand run power stations that burn rice husks.

EARTH : **GLOBAL DECISIONS**

Section 4.5

WASTE NOT ... WANT NOT!

Plants need a continual supply of energy and minerals. Animals obtain their supply of energy and minerals from plants or from plant-eating animals. When an animal or plant dies it still contains stored energy and minerals. Also animal faeces contain stored energy and minerals. Dead bodies and faeces form the food for a group of organisms called decomposers.

microbes. The decaying organism may also have a foul smell that results from a special group of bacteria whose respiration releases hydrogen sulphide. The time for an organism to decay depends on the conditions of the non-living environment. Decay occurs more quickly when these are suitable for fast microbe reproduction. This occurs in warm, damp conditions where microbes can rapidly reproduce. The surroundings also need to have adequate supplies of oxygen for microbe respiration.

Decomposers break down their food, absorbing the minerals and transferring the energy to their own body systems. Earthworms, maggots and beetles scavenge on the dead bodies and faeces. However, bacteria and fungi are the only decomposers that can completely break down energy-containing materials to carbon dioxide and water. These form the most important group of decomposers in any ecosystem as they are the final stage in the energy flow of all food chains.

Soon after an organism has died, fungal spores and bacteria land on it and start to break it down. They also start to reproduce at an amazing rate, spreading all over the dead organism very quickly. The bacteria and fungi digest the dead material by pouring enzymes onto it, gradually turning it into a liquid. Once digested the dead material can be absorbed by the bacteria and fungi and used for respiration. As decay continues, the rotting material may become quite warm. This is caused by the heat energy that is lost from the many respiring

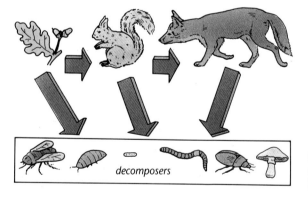

Decomposers take energy-containing materials from all levels of a food chain

N. Discuss

- Discuss the role of microbes in the process of decay.

- Explain why the decay of a dead organism might be faster at different times of the year.

- Explain why it is more likely for whole animals to be found preserved in ice than as fossils.

Energy flows along a food chain and is passed to decomposers whenever an organism dies or produces waste. Eventually it is released from a dead organism or from faeces by decomposers, and spreads into the non-living environment. Energy cannot be recycled. Plant and animals rely on a continuous supply of energy from the Sun to drive their life processes.

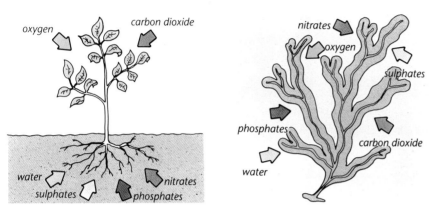

The supply of raw materials to land and water plants

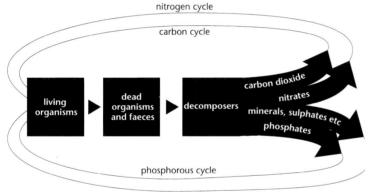

Decomposers' role in cycles

As bacteria and fungi break down dead material and faeces, they release minerals back into the soil. In this way, nitrates, phosphates, sulphates and other minerals are returned to the soil. New plants can absorb and use the minerals that are returned to the soil. In this way, decomposers form an important link in the natural cycles of all ecosystems.

> **O. Work out**
>
> - Use Worksheet GD12 *Energy flow and mineral cycles* to help you produce a summary for this section.

Section 4.6

FUTURE POPULATIONS

Some scientists have estimated that species of animals and plants are now becoming extinct at 25 000 times the natural rate. This could mean that by the year 2000 several species may be lost to our planet. This could have devastating effects for the remaining species. The loss of one plant type can cause the loss of as many as 30 types of insect and other animals that depend on it as a food source.

Wild Ginseng, Panax quinquefolius, *has been used as a healing plant by the people of Asia for thousands of years*

The loss of a single species is a tragedy but the loss of so many may have a significant effect on the future of our planet. Humans rely on other species for food, building, clothing, medicines, furniture, paper, dyes and many more everyday items. Some of these have been used by humans for thousands of years. The plant *Aloe veras,* mentioned in the Bible for its healing properties, is found in many soaps, shampoos and cosmetics today. Beetles were used by the Egyptians to produce dyes, and today we still use the cochineal beetle to supply us with red food colouring. Other species have more recently been found to be useful to humans, particularly in the development of medicines. Snake venom is used in a diluted form to aid blood flow in some people who have circulatory problems. A plant called the rosy periwinkle, found in Madagascar, is used to make two medicines to treat children with leukemia. These new medicines, along with other treatments, have increased the remission rate for children with leukemia to over 80 per cent.

Major drugs derived from plants

Plant	Drug	Use
Amazonian liana	curare	muscle relaxant
autumn crocus	colchicine	antitumor agent
belladonna	atropine	relieves pain
camphor tree	camphor	relieves imflammation
coca	cocaine	local anesthetic
common thyme	thymol	antifungal
foxglove	digitoxin, digitalis	prevents heart attacks
Indian snakeroot	reserpine	prevents hypertension
meadowsweet	salicylic acid*	analgesic
Mexican yam	diosgenin	birth-control pill
mint	menthol	relieves imflammation
Nux-vomica	strychnine	nerve stimulant
opium poppy	codeine, morphine	anelgesic
recured thornapple	scopolamine	sedative
rosy periwinkle	vincristine, vinblastine	antileukemia
tea	caffeine	nerve stimulant
velvet bean	L-dopa	treats Parkinson's disease
white willow	salcin*	analgesic
yellow cinchona	quinine	prevents malaria

* the compound formed from salicylic acid and acetic acid is called acetylsalicylic acid: better known as aspirin.

P. Work out

- Look around the room that you are in and list all the materials that have been made from animals and plants.

Estimated number of species

This table shows the probable number of species and the extent of unidentified biological diverstiy worldwide. If the total number of species if 4.4 milllion, science has described 31% of them. If the total is 33.5 million, only 4% have been described. This is because many species-rich ecosystems like tropical forests contain unexplored and unstudied habitats.

	Number identified	% of estimated total
micro-organisms	5760	3 – 27%
invertebrates	1 020 561	3 – 27%
plants	322 311	67 – 100%
fish	19 056	83 – 100%
birds	9040	94 – 100%
reptiles and amphibians	10 484	90 – 95%
mammals	4000	90 – 95%
Total	1 392 485	

	Number of species	% yet to be identified
low estimate of all species	4 443 644	69%
high estimate of all species	33 526 024	96%

One of the richest areas for plants and animals is the rainforest. A typical patch of 1000 hectare of rainforest contains as many as 1500 species of flowering plants, up to 750 species of tree, about 400 bird species, 150 kinds of butterfly, 100 different types of reptile, around 60 species of amphibian and a large number of various types of insect. In fact there are so many plants and animals in the rainforests that many of them are yet to be identified and named by scientists. Consequently even less is known of the potential use of these unidentified plants and animals. They may perhaps contain important chemicals that could be used for medical purposes, or perhaps be a new food source for humans or for cattle.

There are a large number of products that originate in rainforests:

timber	rubber	gum
latexes	resins	waxes
tannins	steroids	rattans
bamboo	pesticides	nuts and fruit
lubricants	flavourings	dyestuffs
medicines	essential and edible oils	

However, in taking these materials from the rainforests, great areas are being destroyed.

Rainforests are being lost to our planet at a frighteningly high rate. In the last fifty years Latin America has lost 37% of its original rainforest, Asia 42% and Africa 52%. In some cases areas of the rainforest are cleared to produce fruit, nuts, rubber, palm oil, sugar or cotton for export to other countries. In other places rainforest eradication often occurs in stages.

Logging taking place in a rainforest

An amerindian preparing the ground for planting crops

Cattle grazing on cleared land

Logging companies build roads into the rainforest in order to extract timber. Local farmers use these roads to move into the rainforest in search of land. They clear small areas of the forest to grow subsistence crops. To do this they cut down trees and burn them to provide an ash fertiliser for their crops. After only three or four harvests, the land becomes impoverished and difficult to farm. Because of this, the farmers move on to a new area of the rainforest and start to farm that.

Before moving on to a new area of rainforest, some of the farmers sow grass seed. This grows quickly into rough pasture land that the farmers can sell to cattle ranchers for their herds. Cattle ranching is the final stage of the destruction which makes it impossible for rainforest to regenerate in that area.

In recent years many organisations and governments have placed saving the world's rainforests high on their agenda. Many initiatives have been launched to preserve the existing rainforests. However, it is important that these plans include a means of living for the people who live in and near the rainforests that is non-destructive to the rainforest, otherwise the wide scale destruction will continue. Some activities, such as tapping rubber and agroforestry, are generally non-destructive to the rainforest. Also, recent research has shown that tapping rubber and agroforestry achieve much higher cash returns than logging, farming or cattle ranching. Changing people's way of life takes time, but is necessary if there is to be hope for the future of our rainforests.

Q. Plan and present

- Discuss the reasons why rainforests are being destroyed at such a high rate.

- List the ways in which the rainforests are important to humans.

- Plan a talk called *Save our rainforests* that would explain to younger members of your school the importance of our rainforests. You could perhaps present this at an assembly, in lesson time or as an after school session for interested people.

UNIT 5
LAND USE AND MANAGEMENT

Piece of grassland attached to a school being used to encourage wild life

Some of the pictures drawn by the pupils of St Peter's primary school

Section 5.1

BUILDING A POND

It is important that children from a young age are introduced to the wonders of nature. They can begin to understand how ecosystems work. Many primary schools in Britain set up nature areas in their grounds to encourage wild life. Even in large cities birds, insects and other animals can be encouraged to visit or live in these areas. They require suitable plants for food and shelter and not too much disturbance.

St Peter's primary school wanted to set up a school pond and nature area in the school grounds. They were receiving help with this project from a nearby secondary school, who had volunteered to help dig out the pond and also give advice on the pond and nature area. The first stage of the project was a planning stage. Each of the classes at St Peter's were asked to draw their ideas for the pond and nature area. The students at the secondary school looked through all the pictures to get an idea of what the younger children hoped their new area would be like.

A. Work out

- Study the photograph and the children's pictures and advice the children on their project. You might do any of the following:

1. Make a sketch of how you think the pond and and nature area should be constructed so as to include some of the ideas from the children at the school.

2. Make a list of any materials you would like to use and of any plants that might be useful or suitable.

3. Explain the reasons for any materials selected. The list should include how they look, how hardwearing they are or how safe they are for young children to use.

4. Explain why you have selected various plants. Consider their colour or shape, whether they attract wildlife or whether they might provide shade or a home for a particular animal.

5. Study the plans that were made by the secondary school students for the pond and nature area on Worksheet GD13 *Pond plan*.

6. Decide on how their ideas are similar to and different from your plan.

7. Read the instructions that the secondary school students give about the layout of the pond and nature area.

8. Suggest a reason for each of the instructions they give and for their choice of plants.

9. Share your ideas with others in the group and use any important ideas to modify your plans for the pond.

Section 5.2

ASSESSING LANDSCAPES

Brecon in the winter

Britain has many varied landscapes, some of which would be described by many people as 'wild places'. However, the term 'wild place' or 'wilderness' means different things to different people. For some people it suggests a place where the land and surrounding conditions are so savage and desolate that you might risk your life if you entered into it. Indeed, every year lives are lost in mountainous areas such as the Brecon Beacons when people venture out ill equipped to deal with the conditions there. However, many thousands of visitors visit the Brecon Beacons safely and glean considerable enjoyment from walking, climbing and pony-trekking in magnificent surroundings. Many of these visitors think of the area as part of our country where living things can flourish because they find a natural refuge from the influence of industrial Britain.

Another group of people are those who live in or near wild places. The Brecon Beacons has many hill farms where the main type of farming is sheep. These farmers look to the land as a means of earning a living. The farmers and tourists may both speak of a respect and love of the beauty of wild places but the priorities of the farmer and the visitors over land use and management will probably be quite different.

Land use is a controversial topic because people have different ideas on what they feel land should be used for. There is a need to use some of the land for growing crops or grazing animals but many people want to use land for leisure activities. Other groups of people may think it is more important to use the land for building houses or developing industrial estates. However, even within one of these groups ideas may differ greatly. For example, leisure can mean activities as diverse as country walks, windsurfing or visiting theme parks and each of these would mean managing the land in different ways.

The countryside can be used for a wide range of purposes

Major land uses in Britain

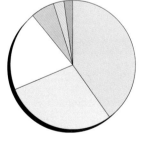

- grass over five years old and rough grazing 40%
- arable and grass leys 29%
- urban, industrial and roads 20%
- broad-leaved trees 3%
- conifers 6%
- farm woodland 2%

B. Work out

- Working in groups of three, discuss how land is used in Britain and how you think the land in the photographs on this page is being used.

- Imagine that part of the land in the photograph of Brecon has been sold to the local authority and they are considering its future use. Each member of the group should take on one of the following roles:

  a neighbouring farmer who wants to expand his farm;

  the secretary of a local rambling group that previously used the land for weekend walks;

  a business executive planning to set up a small electronics factory in the area.

  In your role, write a statement of how you intend to use the land in the future if the local authority gives you permission to use the land.

- Write down all the advantages and disadvantages of your plans but do not show these to the others in the group.

- List all the problems you expect to happen if one of the others is given permission to use the land instead of you.

- Use the ideas you have listed to write an article for your local newspaper on how you feel the land should be used in the future.

- Read each others articles and write a letter for the letter page of the newspaper disagreeing with the article written by one of the other interested parties.

EARTH : **GLOBAL DECISIONS**

C. Investigate

- You are going to carry out a survey of a piece of land to assess its present use and possible future use. For example, the area may be rough grassland that is used for sheep farming, but could be turned into a golf course or a nature trail, or used for rubbish disposal.

- Before going out to look at the site in which you are interested, it can be helpful to have some information on the local history of the land and buildings, and their uses. You might use your library, local authority, local people or local environmental groups to help you to obtain this.

- Decide what criteria you are going to use to assess the site when you go to visit it. Worksheet GD14 *Landscape assessment* might help you to do this.

- Decide how you are going to record the information you collect in your assessment.

- When you visit the site, you need to survey it from several carefully selected viewpoints and so you should decide where these should be. You might select them from an Ordnance Survey map or by taking a quick walk, or cycle ride around the site (depending on the size of the site selected). Mark the viewpoints on a map of the area.

- Carry out the survey assessment and record your observations.

- When you have collected all the information you need, discuss the present use of the land. Decide on a few ideas of how the land could be used in the future. Consider how any changes in land use could affect the people who either live in or use the area of land that you have assessed.

Section 5.3

FARMING LANDSCAPES

Britain covers an area of approximately 23 million hectares. Five thousand years ago most of the land was covered in broad-leaved forest, but today these trees only occupy about 3% of Britain. The majority of the changes in land use came initially from using the land for agriculture and, in more recent times, for towns and roads.

A typical British landscape?

When people think about a typical British landscape, they might imagine green fields, glorious trees and sweet-scented hedgerows. This type of scene is not the natural countryside of Britain, but rather a scene from agricultural Britain. In many people's minds, countryside is equated with agriculture. Following on from this idea, they also believe that the farmer is a type of protector of the land, even a conservationist. While some farmers might have a love of natural beauty, they see their role as growing food and in earning a living from the land.

In recent years, there has been significant changes in farming methods. This has altered the way that farmers use their land which, in turn, has changed the British landscape. An example of this is hedgerow removal.

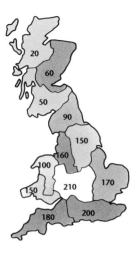

Lengths of hedgerow removed between 1978 and 1984, shown in metres per kilometre square (Source: Institute of Terrestrial Ecology, 1986)

Farmers find it convenient to remove the hedgerows because of the large machinery that modern farms now use for ploughing the land and planting and harvesting crops. Hedgerows and gates make it difficult and time consuming to move the machines in and out of small fields. Removing hedgerows also releases more land for cultivation and the extra crop would mean more profit for the farmer. This, in turn, keeps food prices low for the public.

D. Investigate

- Study Worksheet GD15 *Ploughing times* which shows three fields that need ploughing. Each of the fields covers an area of 12 hectares.

- Decide which field you think could be ploughed the fastest and which would take the longest to plough.

- Working in pairs, one person 'ploughs' each field using a bluntish pencil or thick marker pen as the 'plough', while the partner times how long it takes to plough the field. Try and move the 'plough' at a constant rate and allow time for turning the plough each time you reach an obstacle.

- Discuss the results of the 'ploughing times' for the three fields. Explain why farmers often remove hedgerows and obstacles like trees or ponds from their land.

- Study the map of Britain which gives information on hedgerow removal in Britain between 1947 and 1984.

- List the areas of Britain which have had the greatest changes in their landscape due to hedgerow removal since 1947.

- Study the picture of a British landscape in 1940s, taken in an area that has had large amounts of hedgerows removed in recent years. Draw a picture of how you think the landscape might look today.

A hedgerow provides food and shelter for many animals

A landscape in Britain in the 1940s, when hedges were still very much in evidence

Removing hedgerows may change the look of a landscape but it has other effects too. Hedgerows provide shelter, nesting sites and food for many animals. Removing hedgerows and other modern farming methods have led to a decline in many species over the last fifty years. The following table shows the post 1950 decline in British bird populations associated with farmland habitats (including trees, small woods and cultivated habitats).

Types of bird	Reason for decline
corncrake	modernisation of agriculture, especially mechanised hay and cereal cutting
grey partridge	modernisation of agriculture, reduction of habitat variety
rook	extensive conversion of grassland to arable
corn bunting	uncertain: belongs almost wholly to farmland
yellowhammer	modernisation of agriculture, reduction of habitat variety
cirl bunting	ploughing of old grassland, removal of hedges and scrub
tree pipit	ploughing of old grassland, removal of scrub and woodland
cuckoo	modernisation of agriculture, reduction of habitat variety
yellow wagtail	draining and improvement of wet meadowland
common snipe	draining and improvement of wet meadowland
redshank	draining and improvement of wet meadowland
lapwing	ploughing or improvement of grassland
barn owl	modernisation of agriculture, reduction of habitat variety
little owl	modernisation of agriculture, reduction of habitat variety
long-eared owl	uncertain
sparrowhawk	organochlorine pesticides and reduction of habitat variety
kestrel	organochlorine pesticides and reduction of habitat variety
wryneck	uncertain

Hedgerow removal also has other effects on the land. The soils in which crops are grown can easily be eroded by wind or rain. Removing hedgerows leaves the land more open to erosion and subsequent destruction of farmland, as described in the following article.

Buried by the landslide
By Anthony Tucker

The rate of soil erosion in woodland or heathland is about 0.1 tonnes per hectare per year. It is important to bear in mind that the upward trends in erosion are a direct result of farming techniques, of compaction by heavy machinery and the conversion of unstable soils to arable production.

In East Anglia erosion rates are averaging as much as 18 tonnes per hectare over large areas in the worst years, with local erosion approaching 40 tonnes per hectare. In Bedfordshire, annual losses on sandy loams – primarily from water erosion – have been found to be in the region of 10-45 tonnes a year. Even on chalky soils under cereals, where erosion rates are generally assumed to be trivial, measured losses turned out to be substantial – up to 24 tonnes per hectare in a bad year. In the Midlands the figures are similar.

Water erosion, of course, is a result of particular rainfall patterns interacting with unstable soils. Unstable soils are being increasingly exposed as a Government-driven quick-buck policy for greater cereal production. It turns out that the conditions likely to produce major erosion occur once every three years over most of Britain. On the basis of known soil structure and the present pattern of arable farming in England and Wales, it is now estimated that some 27,000 square kilometres of arable land are unstable. This is about 44 per cent of all arable land and, on the known rates of soil loss, much of this land will be bare and unproductive by the early years of next century.

Real farmers, rather than developers or accountants, are concerned with the long-term renewable productivity of the land. Indeed, land-prices should be substantially based on stability and renewable yield. It may be significant that already in Norfolk, Suffolk and other area arable yield are not reaching predicted levels. The implication is that soil structure is changing under the impact of present techniques. There is a need for coherent independent research.

Source: The Guardian, 28 November 1984

E. Discuss

- Discuss the advantages and disadvantages of hedgerow removal for a farmer.
- Discuss how hedgerow removal is affecting the countryside and your views about farming and the countryside.

Section 5.4

CONSERVING THE COUNTRYSIDE

National Parks of England and Wales

While the majority of British people live in towns, a large proportion of these visit and appreciate the countryside. Britain has a unique and often contrasting collection of landscapes. This is a heritage that should be cared for so that it can be passed on to future generations as part of their heritage.

In 1949, parliament passed The National Parks and Access to the Countryside Act. This act was made in response to public concern over 'the disappearing countryside'. Ten areas became National Parks. Since then the Norfolk Broads have also been designated a National Park. National Parks cover about 10% of the land in Britain. A National Park is defined as *an extensive area of beautiful and relatively wild country in which, for the nation's sake:*

a) the characteristic beauty of the landscape is preserved;

b) access and facilities for public open air enjoyment are amply provided;

c) wildlife, and buildings and places of historical and architectural interest, are suitably protected;

d) established farming use is maintained.

National Parks belong to everyone in Britain, but some people have a particular interest in them because they either live, work or spend their leisure time there. National Park users include:

- the army
- campers and caravanners
- climbers
- potholers and cavers
- farmers
- the Forestry Commission
- water authorities
- boat owners and canoeists
- mining firms
- Electricity Generating Companies
- ramblers
- guides and scouts
- nature reserves
- field centres
- home owners
- anglers

In addition to National Parks, Areas of Outstanding Natural Beauty and Sites of Special Scientific Interest (SSSI) have also been set up across Britain. Many areas have also established country parks and local nature reserves to provide people with access to the countryside. These occupy another 10% of the land in Britain.

National Parks are used by people with a wide range of interests

---

F. Research

- Carry out a survey of people's knowledge and opinions on National Parks.

- Decide how many and what questions you are going to ask them. You will probably get more support to do this if you make sure that it takes only a few minutes to answer the questions and also if the questions are clear and easy to understand. Questions you could ask include:

  Where is the nearest National Park to this town?

  When did you last visit a National Park?

  How many National Parks can you name?

  Do you think there should be more National Parks? If so, which areas?

- Decide how you are going to use the information that you collect and how you are going to present your findings.

EARTH : **GLOBAL DECISIONS**

Different peoples' views about National Parks

Protected areas need to be large as well as being set aside for wildlife. Preserving small patches of land while the surrounding area is used for farming or building does not work. For example, in China twelve nature reserves have been set up in the mountains on the eastern edge of the Tibetan plateau. These protected areas have been established to try and save the giant panda from extinction. Over half of the surviving world population of giant pandas live in these reserves. However, each of the reserves is surrounded by farm land. This makes it difficult, if not impossible, for the pandas to move between the twelve protected areas. Pandas eat huge amounts of only one plant type, the bamboo. They need to feed on this for most of their waking hours. If bamboo is in short supply in a particular area, the pandas there may starve as they cannot reach other areas of food because of the farm land in between. Also, isolating groups of pandas to a particular area results in inbreeding and this can lead to weaker offspring. In the case of the panda, this makes extinction all the more likely. The panda will not be able to survive for long, unless the protected areas are linked by special 'corridors' of suitable habitat and unless the area of all the reserves is enlarged.

G. Discuss

- Discuss the comments made by the people about National Parks.
- Decide which comments you agree with and which you disagree with. Explain your reasons for these decisions.

H. Research

- Carry out a library search to find out information about other animals or plants that have become extinct or likely to become extinct this century.
- Try to find out which events or processes led or are leading to the extinction of the plants or animals you are investigating.
- Pool your information with others in your group.
- Make an exhibition display to show what you have found out to members of the public. Imagine that the exhibition is part of a campaign for more protected areas.

Giant panda

The world's first national park was Yellowstone Park in the USA. It was established in 1872 as a recreation area for the American people. Globally, about 3% of the Earth's surface is protected area either as national parks, nature reserves, wildlife sanctuaries or protected landscape. In 1982, the World National Park Congress stated that 10% of the world's surface should be protected if we are to have any hope of conserving our wildlife.

Section 5.5

FLOODS AND DAMS

For thousands of years, the River Nile has flooded annually. Its waters have irrigated the lands for a few short weeks and have then left behind fertile silt that has helped to fertilise the soil. This is why nearly all Egypt's population lived along the banks of the Nile. By building earth dykes, farmers could trap and store water to irrigate their crops of rice, wheat and maize. The annual flooding also washed away minerals which reduced the fertility of the soil.

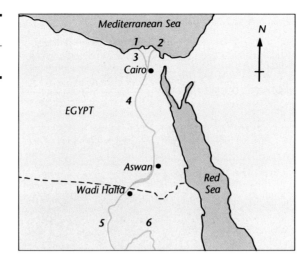

Effects of the River Nile in the early 1960s
1. 19 000 tonnes of sardines each year
2. 32 billion m³ of water 'wasted' by flowing unused into sea
3. Delta built up by annual deposit of silt
4. Basin irrigation – fields flooded each year
5. Water level very variable – high some years, low in drought. Navigation difficult
6. Water snails which carry disease bilharzia washed away in annual flood

1. Discuss

- Make a list of the types of people that benefit from the Nile flooding.
- Discuss with a partner how each of the people in your list benefits from the Nile flooding.
- Suggest and explain any disadvantages of the Nile flooding annually.

Farming on the banks of the River Nile

Some silt was also carried into the Mediterranean sea providing minerals for the marine plant life. This provided food for the marine animals which increased in number, providing a fishing industry. In the early 1960s about 30 000 fishermen worked in the Mediterranean near the Nile outflow, catching sardines and shrimps.

The annual flooding of the Nile also helped reduce the number of people suffering with the intestinal disease bilharzia. This disease is caused by a parasite that lives in the blood vessels of the intestine. For part of its life cycle the bilharzia parasite lives in snails. Many of the snails were washed away when the Nile flooded each year. This reduced the spread of the parasite.

Egypt has a rapidly growing population. The population has doubled over the last 30 years and it is likely to double again within the next 25 years. In the early 1960s, the Egyptian government were very concerned to ensure that enough food could be produced for the Egyptian people. They decided to build a large dam so that they could control the flow of the Nile and use its waters to irrigate the land all year round. This would enable the farmers to grow more food. Modern technology and foreign aid enabled the Egyptians to construct the Aswan High Dam in 1964.

The Aswan dam

EARTH : **GLOBAL DECISIONS** **51**

The Aswan dam holds back and controls the flow of the River Nile. It created a vast reservoir behind it that stretches for some 500 km. This is the second largest artificial lake in the world and was soon used for fishing. As well as providing water for irrigation all year round, some of the water is used to generate electricity in a hydroelectric power station. Another advantage of this scheme is that the Nile today has a more regular flow than it had before the dam was built and this helps boats navigating along it.

Some consequences of building the Aswan dam

J. Work out

- Write a few sentences to explain why the Egyptian government needed a project like building the Aswan dam, in the early 1960s.
- Make a list of the types of people that benefit from the Aswan dam.
- Make a note of the advantages resulting from the Aswan dam.
- Compare these advantages with the advantages and disadvantages of the River Nile flooding, as it did annually before 1964.

Whenever there is a change made to an environment, it affects the natural cycles in that area. It is very difficult to predict the extent to which cycles will be affected because so many factors are involved. The Aswan dam scheme changed a vast area of land and altered the flow of a large river. Within a few years of the dam being constructed, there were noticeable changes taking place in the environment affected by it.

Many environments are stable and continue in much the same way from year to year. However, when people change environments to improve their lifestyles, the stability is upset. This can result in further quite rapid changes to the environment. In the case of the Aswan dam, changes occurred which affected the lives of several groups of people.

K. Discuss

- Study the diagram showing the changes in the hydrology of the area affected by the Aswan dam, and read the comments made by some Egyptians about how the dam has affected their lives. Discuss the reasons for each of the problems and suggest any solutions to them.

L. Work out   EXTENSION

- Write a report summarising your conclusions about the value of the Aswan Dam scheme for the Egyptian people.

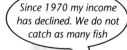

Since 1970 my income has declined. We do not catch as many fish

East Mediterranean fisherman

The lake seems to get shallower every year. This is very worrying

Manager of the Aswan dam

Many more people are ill with bilharzia than when I was a boy

Old Egyptian

My crops keep failing. The soil has become very salty and I have to buy fertilisers these days

Peasant farmer

Section 5.6

ENVIRONMENTAL COLLAPSE

The world's deserts are spreading. If the present spread continues, it could affect the lives of at least 1.2 billion people by the year 2000.

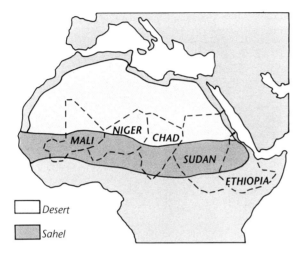

Countries of the Sahel

The area most in danger is the Sahel. This is a band of land lying between the Sahara desert and tropical Africa. Traditionally, African farmers rotated their crops and rested the poorest parts of their land, sometimes leaving it without crops for as long as 20 years.

Traditional African farming

 minerals
 water

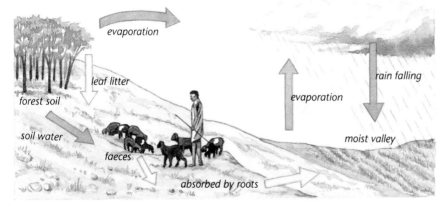

However, the increasing demand for food and firewood in recent times has meant that this method of farming has been abandoned. Forests have been cleared to provide firewood and also more land to grow food on. This has drained the lands of minerals and also has decreased rainfall in the area. Together, these quickly lead to soil erosion and fertile land becomes desert.

The beginnings of a desert

In Ethiopia, the rains began to fail in 1980. By 1985, a crisis situation had been reached. Food stocks were gone. Farm animals had died. Even the grain usually kept back for the following years crop had been eaten. The country was in famine. Many countries sent food and money to aid Ethiopia. Relief camps were set up to try and distribute food and medical care to the Ethiopians. However thousands died and, despite foreign aid, Ethiopia and the other countries in the Sahel are still on the brink of starvation today.

M. Discuss

- Make a list of the items or activities that foreign aid money could be used for in the Sahel.

- Sort these into two lists headed *emergency relief* and *long-term improvement*.

- Decide which item or activity is most important and explain your choice to someone.

In some parts of the world attempts are being made to turn back the deserts. In northern China a wide arc of trees, called the San Bei forest belt, has been planted across the dry hills. It will eventually cover 3.5 million square kilometres and provide shelter for millions of hectares of cropland. Acacia trees have been used to stabilise 60 000 hectares of sand dunes in Rajastan in India. In West Africa, the kad tree has been used to reduce erosion by the sun and wind. Desertification can be stopped but it needs scientific help and money to achieve it.

Desert reclamation in Israel

EARTH : **GLOBAL DECISIONS** 53

UNIT 6
LOCAL INVESTIGATIONS

Methods of capturing small animals

Yagnesh, Meera and Louise decided to carry out a survey of their local area. Meera used quadrats to study the plant populations. Her results table is shown here:

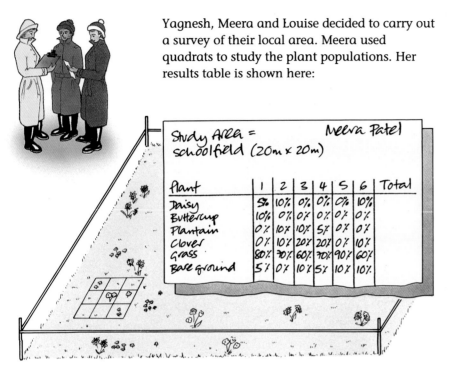

Study Area = schoolfield (20m x 20m)   Meera Patel

Plant	1	2	3	4	5	6	Total
Daisy	5%	10%	0%	0%	0%	10%	
Buttercup	10%	0%	0%	0%	0%	0%	
Plantain	0%	10%	10%	5%	0%	0%	
Clover	0%	10%	20%	20%	0%	10%	
Grass	80%	70%	60%	70%	90%	60%	
Bare ground	5%	0%	10%	5%	10%	10%	

A. Interpret

- Copy and complete Meera's results table.
- Make a barchart to show Meera's results.
- Write a description of what you think the area is like where Meera did her quadrat work.

Louise is interested in the animal populations in the area. From a book search she found several methods of capturing animals.

Pitfall trap

stones and plants to conceal entrance
yoghurt pot
pit

Beating

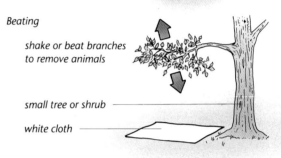

shake or beat branches to remove animals

small tree or shrub
white cloth

Longworth mammal trap

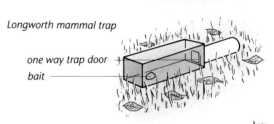

one way trap door
bait

Sweep net

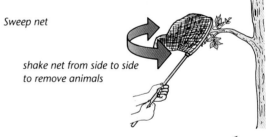

shake net from side to side to remove animals

Pooter

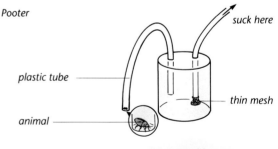

suck here
plastic tube
thin mesh
animal

Night light

EARTH : GLOBAL DECISIONS

Trap 1

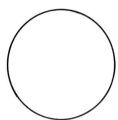

Trap 2

Trap 3

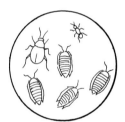

Trap 4

Trap 5

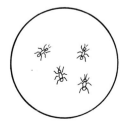

Trap 6

Louise set up six pitfall traps in the area.

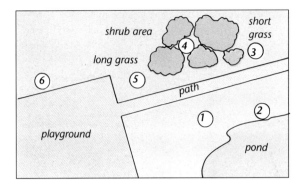

The positions of the pitfall traps

Each day she collected the animals caught in the traps. She noted which types and how many there were of each type before releasing them again. Her results are shown on the left.

B. Work out

- Use Worksheet GD19 *Invertebrate key* to help you to identify the animals Louise caught in the pitfall traps.
- Make a table to show the type and number of animals Louise found.
- Decide on the best way to display these results.
- Study the map showing the sites of the pitfall traps.
- Discuss any ideas that relate the number or type of animal found to the place the animals were caught.
- Write a report of Louise's results and list any future investigations that could be done to further Louise's initial survey.

Louise decided to study woodlice. She noticed that most woodlice were found under stones, amongst piles of fallen leaves or in the crevices between a building and the soil. She used a method called 'mark and recapture' to help her estimate how many woodlice lived in the area. Her results are shown below.

Population estimate
of woodlice – 17th September

Number caught and
marked on Monday = 42

Number caught on
Thursday (ie 2nd sample) = 38

Number of 2nd sample
that were marked = 28 out of
the 38

Population size
(estimate) =

C. Work out

- Use Worksheet GD10 *Mark and recapture* to make an estimate of the woodlice population in the area.

D. Investigate

- Plan and carry out an investigation in the laboratory based on Louise's observation that woodlice are found in particular areas.
- Make a hypothesis based on what you think conditions are like under stones, amongst fallen leaves and in crevices between a building and the soil.
- Decide on the variables you are going to vary, those you will control and what you will record as results.
- Interpret the results and explain how they relate to your initial prediction.

EARTH : GLOBAL DECISIONS

Yagnesh was interested in the effects of the non-living parts of the environment. He took readings of the temperature, wind speed and direction, rainfall and humidity each day. He also carried out tests on the soil to find its components, pH, water content, humus content and air content. Some of his results are shown on the right. These are some of the ideas that the three students came up with:

I found more woodlice where the soil was wetter and slightly acidic. Perhaps they prefer these conditions. We could test it.

There were more plants where there were higher readings of humus. Could an increase in soil humus lead to an increase in plant growth?

% Moisture
Mass of soil + container = 108.6 g
Mass of container = 81.0 g
Mass of soil (before heating) 27.6 g
Mass of soil (after heating) 25.9 g
Mass of water in soil 1.7 g
% water in soil = $\frac{1.7}{27.6} \times 100$ = _____ %

% humus
Mass of soil after burning = 25.0 g
Mass of humus (25.9 - 25.0) = 0.9 g
% humus = $\frac{0.9}{27.6} \times 100$ = _____ %
soil pH = 6

Another interest they had was to try and estimate the affect that humans had on the study area. This is part of a conversation that they had one day:

Meera I think that plants will be affected by people walking on the ground. I am going to look on paths and away from them.

Louise *I am more interested in the effect of our local factory's air pollution on organisms. I am going to look at animals near the factory site and further away.*

Yagnesh Why don't we put our microscope slides smeared with Vaseline on both sites each day?

Louise *I would expect the slides near the factory to collect more dust and dirt, and we would find less animals such as butterflies.*

Yagnesh Yes, this would block their air passages and make it difficult for them to breathe.

Louise *I think the effect of the dirt would be on the plants – they would find it harder to photosynthesise. Poorer plant growth would then affect the butterfly population.*

It would be relatively easy for Yagnesh and Louise to check the air pollution levels using the Vaselined microscope slides and also to gain an estimate of the animals in both areas. However, to test their explanation for gaining these results would be quite a complex and lengthy procedure. Often in Science different explanations are given for a set of results, and this generates discussion and further experimentation which adds to the ever-increasing understanding of Science.

E. Investigate

- Plan and carry out an investigation of a selected area. You may want to use some of the ideas that Meera, Louise and Yagnesh used in their study, or other ideas from earlier units in this book.

- Make a quick survey of the area so that you have an idea of the plants and the animals and the conditions in which they live. You may want to compare it with another area that you have studied to help you do this.

- Make a prediction that links the number or type of organisms with one or more conditions in the non-living part of the environment.

- Decide on which variables you will change, which you will control and what you will record for your results.

- Interpret your results and explain how they link with your initial prediction.

EARTH AND UNIVERSE

Contents

Unit 1	Landshapes	58
Unit 2	Fragile Earth	71
Unit 3	Foundations	85
Unit 4	Journey into space	96
Unit 5	Understanding the solar system	106
Unit 6	Back to the beginning	115

UNIT 1
LANDSHAPES

Section 1.1

WEATHERING

Landforms

From space the Earth looks like a smooth sphere. It is only at, or near, ground level that the land surface displays its variety of landforms such as, mountains, valleys, canyons, plains and coast lines.

The landscape is the result of the processes of weathering, erosion and transportation acting on the geology of an area. The type of landscape produced will also depend on the climate of that area.

Weathering and erosion produce chemicals and rock fragments which can be transported, and later deposited as sediments. The sediments may eventually produce sedimentary rocks. These sedimentary rocks often show evidence of the processes which led to their formation. These same processes, together with the action of plants and animals, are also responsible for producing one of the Earth's most important resources, soil.

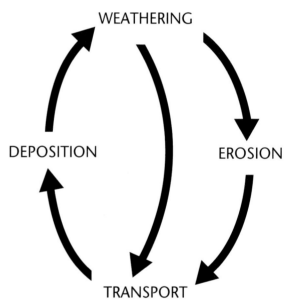

The sedimentary cycle

As soon as a rock comes in contact with the Earth's atmosphere it starts to be broken down or weathered. The type of weathering and the speed at which weathering takes place depends on a number of factors, including rock type, temperature and the amount of water available.

Weathering can be divided into two main types, physical weathering and chemical weathering. The relative importance of each type depends on the climate but, generally speaking, they both operate everywhere.

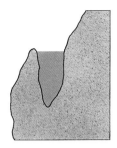
Freeze-thaw weathering causes 'frost shattering'

A
Crack fills with water

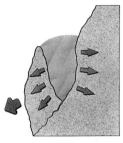
B
As the water freezes it expands exerting pressure which breaks off bits of the rock

Weathering by expansion and contraction

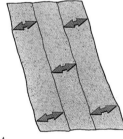
A
Layers of rock expand by day...

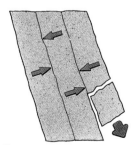
B
... and contract by night causing bits to break off

Physical weathering involves physical forces which break up the rock into smaller fragments. In areas where the temperature regularly rises above 0°C and then falls below 0°C there is a regular freeze-thaw cycle, and 'frost shattering' can occur. When water freezes it expands by about 8%. When the water in a pipe freezes it expands and can rupture the pipe. Similarly any water which has found its way into cracks, or into the spaces between minerals or grains in a rock, exerts a force when it freezes. Regular freeze-thaw can cause the rock to break up into sharp-cornered fragments. These fragments can be of a variety of sizes and are often seen forming scree at the foot of mountain slopes.

Stresses can also be built up in a rock by changes in temperature. Minerals in a rock expand when they are heated, and contract when they are cooled. Different minerals expand and contract at different rates. In areas where there are daily extremes of temperature the outer layers on rock are prone to breaking up.

A. Investigate

- Devise your own investigation into the effect of physical weathering on different types of rock.
- Discuss your ideas with your partner and produce a plan to show your teacher. Your work should include:

  a hypothesis about which rock(s) will be most/least effected and why;

  a discussion about how well your investigation models the effects in nature;

  a description about the places where these different types of physical weathering are most likely to occur.

The chemical reactions involved in chemical weathering rely on the dissolving action of water and the action of substances dissolved in water. In addition to the presence of water, temperature is also an important factor. For every 10°C rise in temperature there is a doubling of the rate of reaction. Chemical weathering takes place most rapidly in hot and wet areas, but even in desert regions there is enough water present for some chemical weathering to take place.

The effects of chemical weathering will depend on the minerals contained in a rock. Minerals which contain iron compounds react with oxygen from the atmosphere and, in the presence of water, become oxidised. Oxidation causes these minerals to break down chemically. Oxidation can affect chemicals other than iron.

$$4Fe + 3O_2 \longrightarrow 2Fe_2O_3$$
$$\text{iron} \quad \text{oxygen} \longrightarrow \text{iron oxide}$$

Some minerals take up water molecules into their crystal structure. This is called hydration. The minerals expand, leading to stress and causing the rock to break up. Hydrolysis of the mineral feldspar in the granites of South West England produced a clay mineral called kaolinite, or china clay, which is used in the manufacture of china ware and porcelain.

A china clay pit

Rain water is naturally slightly acidic. Rain water reacts will carbon dioxide in the atmosphere to form carbonic acid (H_2CO_3).

This acidic water can then react with carbonate minerals, particularly the mineral calcite from which limestones are formed.

$$2CaCO_3 + H_2CO_3 \longrightarrow 2Ca(HCO_3)_2$$

calcium carbonate   carbonic acid   calcium hydrogencarbonate

Calcium hydrogencarbonate dissolves easily, and is removed in solution. It is this solubility of limestones which accounts for spectacular underground caverns and other landform features found in limestone areas. Industrial pollutants such as sulphur dioxide form sulphuric acid in rain water (acid rain), and add to the effects of chemical weathering of limestones not just on the landscape but on buildings as well.

The mineral quartz is resistant to chemical weathering and passes through the sedimentary cycle relatively unchanged. Not surprisingly quartz makes up much of the sediments in rivers and beaches, deserts and many types of soil.

Chemical weathering of limestone

The activities of plants and animals can increase the effects of physical and chemical weathering. Plant roots force open cracks in the rocks. The burrowing activities of worms allow air and water through the soil to weather the rock below. Rotting plant material produces humic acid which can react with minerals in a rock and help to break it down.

Limestones weather most rapidly, followed by sandstones, igneous rocks such as granites and metamorphic rocks such a quartzites.

Rock type	Rate of weathering (cubic metres/square kilometre/year)
quartzite	1
granite	3
sandstone	15
limestone	75

B. Investigate

- Construction companies have scientists working for them who test different types of materials to decide if they are suitable for building. You have been asked to provide a report about the different types of building stone which have been suggested for the sculptures for a new church.

  1. Devise a test(s) for these building stones to see how they will stand up to chemical weathering. Think carefully about what variables you will need to consider.
  2. Carry out the test(s).
  3. Try to quantify your results, and then present them as a written report for the chief architect.

C. Research

- Find out about the effects of weathering in your local area. This could be on different rock types in the area or on buildings. Many buildings will be made of local building stones such as limestones and sandstones. Some buildings such as banks are often faced with igneous rocks.

  You could create a rock trail around your area, showing the use of different rock types and any special features such as weathering.

Geology on the doorstep

Section 1.2

TRANSPORT, EROSION AND DEPOSITION

The products of weathering – rock fragments, mineral fragments and chemicals – are usually transported (carried away) from the site of weathering. These materials may be moved by water, ice or wind. During movement the products of weathering continue to be broken down chemically and physically. The type of transportation and the time spent being transported will determine in what form these materials are finally deposited as sediment.

Water, wind and ice can produce more sediments by wearing away the rocks over which they travel. This process is called erosion. Some features of the landscape, such as valleys, are formed mainly by erosion; other features, such as beaches, largely by deposition.

Two things can happen to the material which has been transported. The action of wind and water can sort the material. The further the material is carried the more it becomes sorted out into different sizes such as pebbles or sand. Large particles such as boulders and pebbles can only be laid down by fast currents. Sand and mud can be deposited by slower currents. In addition, particles become more rounded the further they are transported as their corners become worn away.

Angular rock fragments

Well-rounded rock fragments

Landforms are produced by erosion and deposition

D. Investigate **EXTENSION**

- Investigate the effects of abrasion on the production of smaller, more rounded material by shaking cubes of plaster of Paris in a plastic container or a cardboard tube. You will need to consider:

  the change in shape of the material;

  the amount of time they are shaken;

  the change in size of the material;

  the change in mass of the material.

- Try to quantify your results and include graphs in your report.

The rate at which a sediment is deposited varies greatly. In a flood a lot of sediment can be deposited very quickly, at other times the laying down of material can be extremely slow.

Water

Rivers can transport sediment in three ways. Larger materials such as pebbles and sand slide, roll or bounce along the river bed. This can only happen with a fast flowing current. Finer material such as clay particles are suspended in the water and can be carried by slower currents. Dissolved chemicals are also carried in the water.

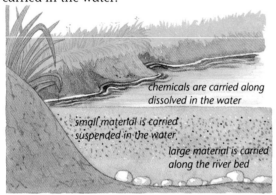

How sediment is transported by a river

A river's power to erode comes from the combined effects of the force of the moving water itself, the grinding down of the bedrock by the movement of sand and gravel, and chemical action as the water reacts with rocks over which it is flowing. These factors also cause the sand and gravel particles to become smaller and more rounded.

When a river loses energy, by slowing down or drying up, some of the material it is transporting will be deposited. This can happen at a number of sites in the course of a river.

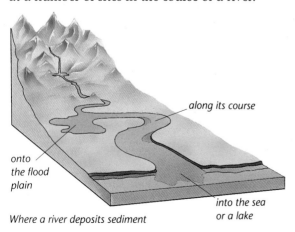

Where a river deposits sediment

Where a river bends or meanders, erosion takes place on the outside of the bend where the water flow is fastest, and deposition takes place on the inside where the water flows more slowly.

Erosion and deposition where a river bends

Deposition also occurs where lowland rivers overflow during periods of flood. As the water spreads out to flood an area it loses energy and deposits silt and mud over the flood plain. This fine sediment is called alluvium and can produce an excellent soil.

Most of the sediment transported by rivers is deposited in the sea. When a river flows into the sea (or a lake) it loses energy, and sediment is deposited. If the tides can remove the sediment faster than it is deposited, an estuary will form. If, on the other hand, the sediment is deposited more rapidly than it can be removed then a delta is produced. The Mississippi delta has developed because two million tonnes of sediment are being deposited each day.

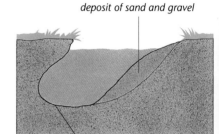

The Mississippi delta, as seen from the shuttle Challenger in 1985

E. Investigate

- The apparatus shown in the diagram can be used to investigate a variety of factors associated with transport, erosion and deposition by rivers. For example:

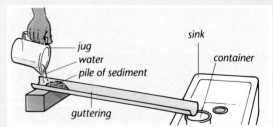

the effect of changing the rate of water flow on sand and gravel;

the way material is moved by water currents (does it float, roll along or slip on the bottom);

the formation of a delta.

Choose one of these factors to investigate. Be prepared to report your findings to the rest of the class.

The waves, tides and currents continue to transport and deposit sediments after they have been deposited in the oceans. Waves are also powerful agents of erosion, producing coastal land forms such as cliffs, bays and headlands. Coastal erosion also provides more sediments.

The constant movement of sediments on a beach by waves means that the sand and pebbles tend to be well rounded and that weaker rocks and minerals are broken down. Beach sands are mostly made up of quartz grains, and pebbles are often made of resistant rock types such as granites. Wave action also sorts sediments, so that some beaches are made only of pebbles and other beaches are made only of sand.

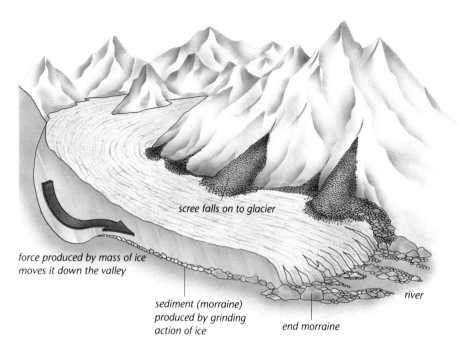

The main features of a valley glacier

A beach showing changes in grain size

F. Think about

- Explain how the following might occur:

    a beach made up only of pebbles;

    a beach made up only of sand;

    a rocky shore line with no beach.

Away from the coastline there will be deposits of gravel, sand and mud, which are eroded and deposited by ocean currents.

The deepest parts of the world's oceans are too far away from land for sediment to reach them. The most common deep-sea deposit is ooze (mud) made from the skeletons of microscopic animals and plants that form plankton at the sea surface. These fall to the ocean bed when they die forming this mud.

Ice

Glaciers form where the temperatures are low enough for snow to last all year. As the snow gets deeper and deeper the lower levels are compressed forming ice. This ice begins to move down slopes under the influence of gravity. As more snow is added a glacier is formed.

Glaciers move relatively slowly, only about one metre per day, but have a great deal of erosive power. For the same volume, ice has 20 times the erosive power of water. Erosion occurs as rock material is frozen into the base of the glacier and is pulled away as the glacier moves. The debris in the base of ice allows abrasion of the rocks below. This grinding action produces a very fine clay-like sediment called rock flour. This in turn also has an abrasive action. The material eroded by the ice ranges in size from clay to huge boulders. Much of the fragmentary material has very sharp corners and these sediments of different sizes are all mixed up together.

When a glacier eventually melts, it dumps all its load together as a poorly sorted mixture of clay and boulders called till. As the ice melts the water flows away in rivers which carry the sand and clay away and deposit them elsewhere.

Glacial till

EARTH : **EARTH AND UNIVERSE** **63**

Sand dunes in the Sahara desert

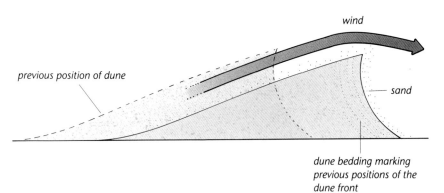

Dune migration and dune bedding

Sandstone cliffs eroded by the wind

Wind

In dry areas, where there are few plants protecting the sediment, wind can easily cause erosion. As with flowing water, the greater the flow of wind the larger the sediment which can be transported. Usually wind cannot move sediments any larger in size than dust and sand.

As wind blows sand grains along they collide with one another, rolling and bouncing. The impact between the sand grains causes the grains to become very well rounded. Not surprisingly only the most resistant materials survive and the majority of sand is composed of quartz.

Wind can deposit sand as broad sheets or as dunes. If there is plenty of sand available dunes move with the wind. Sand grains are moved up the gentle back slope of the dune by the wind. They then fall down the steep front slope. Thus the dune moves downwind. If a dune was cut through the pattern of the layers would be seen. This pattern is called dune bedding.

Sand dunes can also develop at the back of sandy beaches, the result of strong onshore breezes. Wind has also been responsible for depositing silt. This fertile sediment covers parts of Europe, Asia and the USA and was originally the rock flour produced by glacial erosion. The wind transported and then deposited it far from where it was first deposited.

G. Work out

- Use Worksheet EU1 *Changing outline* to show:

  three place where erosion is very active;

  three places where sediment is being transported;

  three places where you think deposits of sediment are building up.

- Describe how erosion, transport and deposition might be different in each of these areas at different times of the year.

Section 1.3

SOIL

The formation of soil

1 Weathering begins to break rocks down

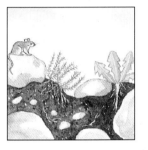
2 Plants and animals colonise the area helping the breakdown

3 A soil develops as dead plant and animals add nutrients

Soil is formed by the action of plants, animals and weathering on rocks. Time is also a factor, it takes many years for a soil to develop. Different types of soils develop in different climates and on different rock types. Upland soils tend to be thinner than those in lowland areas. Plants have the ability to hold a soil together and prevent its erosion.

There are four main components to soil:

mineral content (clay or sand)	50–60%
organic matter	up to 10%
air	15–25%
water	25–35%

The organisms present in the soil are also important. Soil begins to form as the processes of weathering break down the rock. If this material is not transported away it begins to accumulate. Plants and animals become established in the new soil and they help the weathering of the rock. Their dead remains add to the nutrients in the soil attracting more plants and animals. Eventually, over a long period of time, a proper soil will develop. A slice taken through the soil down to the parent rock is called a soil profile.

The mineral content of the soil comes from the weathering of the rock. Different types of rocks weather differently, to give different types of mineral grains. In general a soil has varying proportions of sand (quartz) and clay minerals.

Soil profile

The size of the mineral grains will effect how much water and nutrients can be held in the soil.

An ideal soil would contain equal proportions of sand and clay, giving a soil which will hold water and nutrients but which will not become waterlogged.

The organic part of a soil is made up of dead plant material. Organisms, such as worms, which break down this organic material also do an important job in allowing air and water into the soil, helping drainage.

Air, along with water, can occupy the spaces between the grains in a soil. The oxygen in the air is vital for the animals living in the soil. In a waterlogged soil there is no oxygen so that acids accumulate. This lowers the pH of the soil, making it very acidic.

	100% sand	100% clay
particle size	0.06 – 2 mm	less than 0.002 mm
structure	none	forms lumps when wet, goes hard and cracks when dry
gaps between grains	large spaces between grains	small spaces between grains
water holding	poor due to good drainage	good but easily waterlogged
nutrient holding	low due to good drainage	high (water cannot carry nutrients away)

H. Research and test

- Compare and contrast the soil from two different localities in your area. You may need to do some research into the geology of the area to select suitable sites for investigation. Avoid collecting soil from agricultural land. (Why?)

  If it is possible visit the sites and dig pits to look at and compare the soil profiles.

- Use Worksheet EU2 *Soil tests* to help you to determine the water, dead plant, and air content of the two soil samples, their pHs and the amount of solid material they contain. You may also like to investigate differences in the organisms found in the soil.

Section 1.4

FROM SEDIMENT TO ROCK

When sediment is deposited by water, wind or ice the sediment is loose. As the sediment is buried by more sediment it becomes pressed together (compacted). Eventually the pressure may be high enough for grains to fuse and a solid rock to be formed. Sometimes a mineral such as calcium carbonate may crystallise out of the water that flow through the gaps between the sediment grains. This form a cement which binds the grains together into a solid rock. The process of turning sediment into rock is called lithification.

Clastic sedimentary rocks are classified on the size of their grains. Coarse-grained clastic sedimentary rocks contain a majority of grains over 2 mm in diameter. Medium grained are between 0.05 mm and 2 mm, and fine grained are less than 0.05 mm.

Fragment		Diameter	Sedimentary rock
coarse	pebble	4 – 64 mm	conglomerate
	gravel	2 – 4 mm	
medium	sand	0.05 – 2 mm	sandstone
fine	silt	0.005 – 0.05 mm	siltstone
	clay	less than 0.005 mm	mudstone

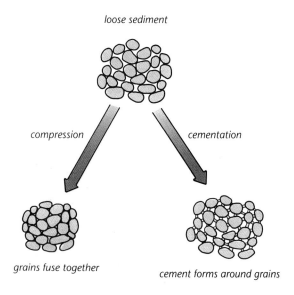

Turning sediment into rock

Sedimentary rocks are formed by physical, chemical and biological processes. There are three main types of sedimentary rocks:

clastic sedimentary rocks are those formed by physical processes from material such as clay particles, silt, sand and pebbles;

chemical sedimentary rocks are formed when chemicals dissolved in the water are deposited by crystallising from solution;

biologically formed rocks include fossil-containing (fossiliferous) limestones and coal.

Some rocks are formed by a combination of processes, such as a limestone formed from physically deposited shell fragments.

1 Conglomerate
2 Sandstone
3 Mudstone

I. Observe and record

- Use Worksheet EU3 *Clastic sedimentary rocks* to help you to find out about the sedimentary rocks you have been provided with.

Limestones usually form in the oceans. Limestones are made up of calcium carbonate in the form of the mineral calcite. Many marine organisms extract calcium carbonate form the sea water to produce their shells. Limestones may form from the accumulation of this shell material.

Shelly limestones are composed of the fragmentary remains of shells and skeletons from a range of different types of marine organisms. They must have formed in clear sediment-free waters. Chalk is made up from the skeletal remains of a tiny plant plankton called a coccolith, which is only a few thousandths of a millimetre in diameter. Billions of these would have to have accumulated to have formed the great thickness of chalk found in southern Britain.

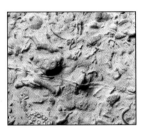
Shelly limestone

Chalk

Oolitic limestone

Another type of limestone particularly common in southern Britain is oolitic limestone. This is formed were silt grains or shell fragments are rolled by currents over lime mud. A layer of mud sticks to the fragment as it is rolled around, and successive layers develop. Tiny spheres, called ooliths, less than a millimetre in diameter are formed. These eventually become cemented together to form limestone. Coral reefs can be found in sedimentary rocks, preserved as reef limestone. Reef limestone will only form under very special conditions. Coral reefs only develop in warm, clear, tropical seas at depths no greater than 20 metres. At these depths there is plenty of light and current action to ensure a food and oxygen supply and the removal of any sediment.

> ### J. Work out
>
> - Use Worksheet EU4 *Limestones* to help you to learn more about limestones.

Under conditions of strong evaporation some of the chemicals found in solution in sea water combine and are precipitated as salts. These salts include halite (NaCl) and gypsum ($CaSO_4 \cdot 2H_2O$). In Britain there are deposits of halite and gypsum tens of metres thick which formed in shallow seas under hot dry conditions millions of years ago.

> ### K. Investigate
>
> - Evaporate some sea water and work out how much salt there is dissolved in it.
>
> - Calculate how much sea water you would need to evaporate to produce 1 kg of salts.
>
>   How would it be possible to produce very thick salt deposits in a shallow sea?

Coal and oil are formed by the decay of plant material; coal by the decay of land plants under swampy conditions and oil by the decay of microscopic marine plant plankton in the oceans. They are two important sources of fuel.

Opencast mining of coal (top)
Oil spouting from a drill pipe (bottom)

For a coal deposit to form there needs to be lots of vegetation, as in hot wet tropical climates. There also need to be swampy conditions. In such conditions there is little oxygen available for bacteria to break down the plant material. As the plant material becomes buried under more plant material and sediments it becomes compressed, and as it is buried deeper the temperature rises. Chemical changes then take place over millions of years to turn the cellulose of the plant material into carbon.

$$C_6H_{10}O_5 \longrightarrow CO_2 + 3H_2O + CH_4 + 4C$$

cellulose   carbon   water   methane   carbon
                dioxide

The longer and deeper the burial the more carbon is produced. In terms of fuel quality this means a better fuel.

Oil is produced in a similar way. Billions of dead plankton accumulate in the muds of the ocean floor where there is little oxygen. The plankton are buried, and the deeper they are buried the higher the temperature. The chemicals in the plankton, particularly substances like chlorophyll, are broken down into the black sticky liquid called oil.

L. Work out

- Use your understanding of how sedimentary rocks are formed to write a description of how, and under what conditions, the sedimentary rocks in the diagram on Worksheet EU5 *History in the rocks* were formed.

Section 1.5

TIME FOR A CHANGE

One of the characteristic features of sedimentary rocks is that they are deposited in layers. Each layer is called a bed and is separated from the next bed by a bedding plane. The bed deposited first, at the bottom, must be the oldest and the bed deposited last, at the top, must be the youngest. This is called the 'principle of superposition'. In a sequence of sedimentary rocks the youngest will always be at the top.

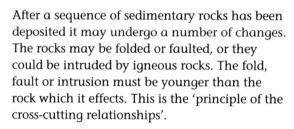

Horizontally-bedded sequence of rocks

After a sequence of sedimentary rocks has been deposited it may undergo a number of changes. The rocks may be folded or faulted, or they could be intruded by igneous rocks. The fold, fault or intrusion must be younger than the rock which it effects. This is the 'principle of the cross-cutting relationships'.

By using these principles it is possible to work out the sequence of geological events in an area without having to give exact dates for the events. This is called 'relative dating'.

The fault cuts across the beds and must be younger

M. Work out

- Collect Worksheet EU6 *Relative dating* which shows a section through a quarry. A number of events have happened to produce the geology shown in the quarry. Use the principles of superposition and cross-cutting relationships to work out the geological events in sequence.

Using relative dating techniques geologists in the eighteenth century were able to work out the order of geological events for the sedimentary rocks of Britain. Fossils played an important part in this process, as rocks of different ages contain different fossils. The geologists recognised patterns in the rocks and used these to work out a sequence called a geological column. They further divided the column into three eras based on the changes in the types of plants and animals found as fossils.

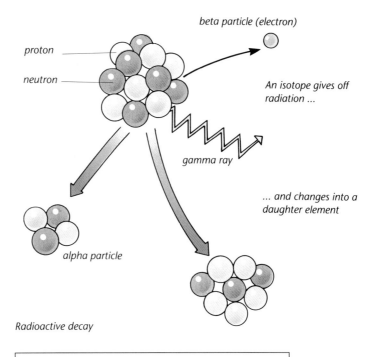

Radioactive decay

	Period	Era
youngest ↑	Quaternary Tertiary	Cainozoic (recent life)
	Cretaceous Jurassic Triassic	Mesozoic (middle life)
oldest ↓	Permian Carboniferous Devonian Silurian Ordovician Cambrian	Palaeizoic (ancient life)

Being able to give a date to an event is called absolute dating. To find the absolute age of a rock geologists use radiometric dating. This involves measuring very small amounts of radioactive atoms. Radioactive elements such as rubidium (Rb) and an isotope of potassium(K) called K-40 occur mostly in minerals in igneous and metamorphic rocks. These elements are unstable and start to decay from the moment they are formed. The original atom, the parent atom, changes into another isotope, the daughter isotope. This radioactive decay gives off energy as radiation.

The time take for half of the parent isotope to change into the daughter isotope is called the half-life.

The half-life for various types of radioactive decay are known from measurements in the laboratory. Different isotopes are used for dating rocks of different ages and have been used to date the oldest rocks on the Earth. These techniques can also be used to date archaeological finds.

N. Observe and record

- A cube of Plasticine represents all the parent isotope present in a mineral when it was first formed. Assume the half-life for this isotope is one million years.

  Cut the cube in half and separate the parent from the daughter.

- Repeat this process for four more half-lives. At each stage calculate the percentage of daughter and parent isotopes.

- Use your data to plot a graph of percentage of daughter atoms against time.

Parent isotope	Daughter isotope	Half-life	Dating range
carbon-14	nitrogen-14	6000 years	100–65 000 years
uranium-235	lead-207	700 million years	10–4600 million years
rubidium-87	strontium-87	50 000 million years	10–4600 million years
potassium-40	argon-40	12 000 million years	0.1–4600 million years

To calculate the age of the rock you multiply the number of half-lives, by the known half-life for that isotope. A sample containing 50% of lead-207 and 50% of uranium-235 will have undergone one half-life and will be 700 million years old.

Geological map of Britain and Ireland
The map shows sedimentary rocks classified according to their age of deposition and igneous rocks according to their mode of origin. The colours are those of the international for geological maps. Figures indicate age in millions of years.

Sedimentary rocks
Cenozoic
  Tertiary Mainly clays and sands up to 65
Mesozoic
  Cretaceous Mainly chalk, clays and sands 65-140
  Jurassic Mainly limestones and clays 140-195
  Triassic Sandstones and conglomerates 195-230
Palaeozoic
  Permian Mainly magnesian limestones and sandstones 230-280
  Carboniferous Limestones, sandstones, shales and coal seams 280-345
  Devonian Sandstones, shales, conglomerates, slates and limestones 345-395
  Silurian Shales, mudstones, some limestones 395-445
  Ordovician Mainly shales and mudstones; limestone in Scotland 445-510
  Cambrian Mainly shales, slate and sandstones; limestone in Scotland 510-570
  Late Precambrian Mainly sandstones, conglomerates and siltstones 600-1000

Metamorphic rocks
  Mainly schists and gneisses 500-1000
  Mainly gneisses 1500-3000

Igneous rocks
  Intrusive: Mainly granite, gabbro and dolerite
  Volcanic: Mainly basalt, rhyolite, andesite and ash

O. Work out EXTENSION

1 An igneous rock is found to contain 75% of the daughter isotope lead-207. How old is the rock?

2 The diagram shows a sedimentary rock sequence containing two lava flows.

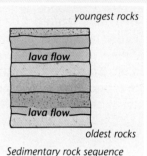
Sedimentary rock sequence

The lower lava flow contains 12.5% of the daughter isotope lead-207. The upper lava flow contains 25% of the same daughter isotope. What does this tell you about the age of the sediment?

The geological column shows the absolute ages of the of the geological periods together with information about the sediment deposited during each period. Most metamorphic rocks in Britain are very old and were formed around 1000 million years ago. Igneous activity, particularly volcanic activity has been quite common throughout the geological history of the British Isles.

P. Work out

1 On the outline map of the United Kingdom on Worksheet EU7 *Geology of Britain* show in red the areas of igneous rocks, in blue the areas of metamorphic rocks and in green the areas of sedimentary values.

2 Is there a pattern to the distribution of the different rock types in Britain:

  a by age;
  b by area?

The processes of weathering, erosion and deposition can effect the landscape. However, these processes only modify a landscape which has been already produced by forces in the Earth which build mountains, form volcanoes, and which cause continents to move.

UNIT 2
FRAGILE EARTH

Section 2.1
HOT ROCKS

Sheets of red hot molten lava from the eruption of Mt Etna, Italy

A year rarely goes by without a major volcanic eruption or an earthquake making the headlines. These rapid local changes to the Earth can cause immense damage and loss of life. To us volcanic eruptions and earthquakes are often seen as short lived destructive events. However, it is Earth forces like these which have been responsible for the formation of new crust, its modification and movement, for thousands of millions of years.

A volcano marks the point on the Earth's crust where molten rock (magma) reaches the surface. Magma is liquid rock containing dissolved gases and water. Magama forms deep within the Earth, where there are high temperatures and pressures. At depths of 200 km rock melts at a temperature of 650°C, but as the magma rises it cools. As it cools, crystals of different minerals form until finally solid rock is produced. Rocks produced by the cooling of magma are called igneous rocks. They are usually crystalline, which means that they are made up of interlocking crystals of different minerals.

A. Think about

- With the top firmly on the lemonade bottle the lemonade stays inside because it is held under pressure. When the top is removed the pressure is released and the lemonade erupts. How do you think this model might help to explain how magma reaches the surface?

  What other factor might be important?

In the Earth between the depths of 50 km and 250 km about 5% of the rock is molten. If the pressure in the overlying rocks is great enough, this molten rock is trapped. If the pressure is released, due to a weakness in the overlying rocks, the magma can rise up and may reach the surface.

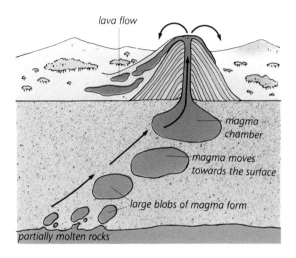

Movement of magma through the crust

There are two main groups of igneous rocks. Intrusive igneous rocks are those which have crystallised before the magma reached the surface. These form features such as batholiths, dykes and sills within the crust. Granite is an example of an intrusive igneous rock. Extrusive igneous rocks are those which have crystallised on the Earth's surface. When magma reaches the surface it is called lava. Basalt is an example of extrusive igneous rock. Lava flows can form both on land and under the oceans.

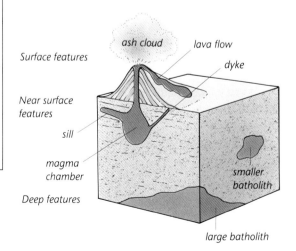

Intrusive and extrusive igneous features

EARTH : **EARTH AND UNIVERSE**

Two properties are used to identify and give a name to an igneous rock: the size of the crystals, which gives a clue about the cooling history of the rock, and the types of minerals present.

Granite

Rhyolite

The igneous rocks in the photographs above both contain the same minerals; they have the same chemical compostion, but they have different sized crystals. The granite has large, easy-to-see crystals but the crystals in the rhyolite are very difficult to see. What can the size of the crystals tell us about the history of these two rocks?

The size of crystals is related to the cooling history of an igneous rock. The larger the crystals the more slowly the rock cooled; the smaller the crystals the more rapidly it cooled. Coarsely crystalline rocks like granite are intrusive and cooled very slowly; finely crystalline rocks like rhyolite are extrusive, forming lava that crystallised quickly.

> ### B. Observe and record
>
> - Each of the samples of igneous rocks which you have been given has differents sized crystals or a pattern of different sized crystals. Use salol or a similar substance to model the size and pattern of the crystals in these specimens.
> - The two photographs on the right show an igneous rock which has a number of large crystals surrounded by many smaller ones and another which has no crystals at all.
>
>   Explain how these igneous rocks might have been formed from liquid magma by cooling.

Porphyritic granite

Obsidian (volcanic glass)

The samples of granite and gabbro in the photographs have a similar grain size. They are both coarsely crystalline rocks and therefore intrusive. They look different because the minerals they contain are different. The granite contains more lighter coloured minerals, and the gabbro more darker coloured minerals. The most common minerals found in igneous rocks are quartz and feldspar (both light coloured) and hornblende and augite (both dark coloured). Granite contains 90% quartz and feldspar; gabbro contains no quartz and about 50% augite.

Gabbro

Quartz is a form of silica, so igneous rocks which contain a lot of quartz are often called silicic igneous rocks. Those which contain no quartz are called basic. Rocks such as diorite, which contain some quartz, are referred to as intermediate rocks.

> ### C. Work out
>
> - Copy out and complete the following table.
>
> | SILICA CONTENT | high | medium | low |
> |---|---|---|---|
> | TYPE | silicic | intermediate | basic |
> | COLOUR | | | |
> | CRYSTAL SIZE Coarse > 5mm | | — | |
> | Medium 1-5mm | — | — | dolerite |
> | Fine < 1mm | | andesite | |

Section 2.2

VOLCANOES

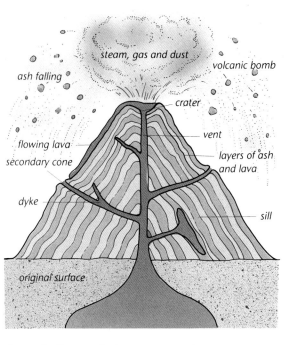

Composite diagram of volcano

> ### D. Investigate
>
> - You are going to investigate the factors that effect the viscocity of a lava, using treacle as your model and any of the following apparatus:
>
> | | | |
> |---|---|---|
> | sand | water | drinking straw |
> | boiling tube | thermometer | stop watch |
> | tile | beaker | Bunsen burner |
> | tripod | safety mat | safety glasses |
>
> First discuss with your partner what factors might effect viscosity. Predict which of these factors will have the most effect on the viscosty of the treacle and explain your predictions.
>
> - Plan how you will carry out your investigation, paying particular attention to fair testing and on how you will make measurements.
>
> - Discuss your plan with your teacher and then carry out the investigation.

Volcanoes form major features on the Earth's surface. There are two basic types of volcano; the central type which has a round central opening, and the fissure type where the opening is a long, crack-like feature. Volcanoes on the continents tend to be of the central type and usually produce andesite lava. Volcanoes under the oceans can be of the central type, many of the islands of the Pacific are the tops of such volcanoes, but most of the eruptions in the oceans are of the fissure type.

There are thousands of volcanoes on Earth. About 500 have erupted within living memory and are described as active. Many more are in a so called resting state, or dormant, others are thought to be extinct. Some volcanoes erupt quietly; others erupt very explosively.

The amount of silica (quartz) present in a magma has an effect on how runny it is when it reaches the surface. The greater the amount of silica (quartz) the thicker or more viscous a lava is. This, in turn, has an effect on the type of eruption there is at the surface. Other factors such as the amount of dissolved gas and water also effect the viscosity of a lava.

The shape of a volcanic cone largely depends on how runny the lava is. Basalt (low in silica) forms a free-flowing runny lava which can flow great distances before it solidifies. It forms large broad cones like that of Mauna Loa. Lavas such as andesite and rhyolite have a higher silica content. These types of lava flow more slowly and much shorter distances before solidifying. Narrow steep-sided cones like that of Mount Fuji-san are produced by this type of lava. Some volcanoes produce very little lava but large quantites of ash and other fragmentary material and produce ash cones.

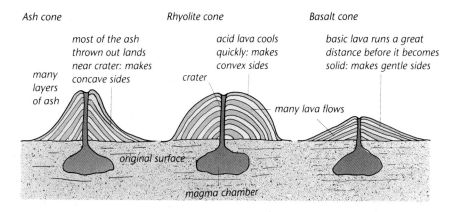

Types of volcanic cones

Rhyolite lavas are often so thick that they solidify in the vent forming a 'plug'. This causes immense pressure to build up. When the volcano does erupt it is extremely violent, and the effects can be devastating. Often the plug is stronger than the side of the volcano, and the volcano explodes sideways sending out a glowing cloud of very hot gas and ash at speeds of over 100 km/h. All 30 000 inhabitants of Saint Pierre (Martinique, West Indies) were killed within seconds when Mount Pelee erupted in this way in 1902. When Mount St Helens (Washington State, USA) erupted in 1980 it produced two cubic kilometres of ash, killed 60 people and caused damage estimated at two billion dollars. The cloud of hot gas and ash produced by mount St Helens was moving so fast that the occupants of a car travelling at 70 mph were killed. We know this because the people survived in a car, which overtook them by travelling at 100 mph.

Occasionally eruptions are so explosive that volcanoes literally blow their top. When Krakatau (near Java) erupted in 1883, the explosion was heard 4000 km away, produced 18 km^3 of ash and left a very big hole which was rapidly filled by the ocean.

Volcanoes are not just found anywhere on Earth. There is a definite pattern to their distribution on the Earth's surface. Active volcanoes are confined to narrow volcanic zones.

The eruption of Mount St Helens, USA

E. Think about

- Transfer the information from the diagram on to the map of the world on Worksheet EU8.
- Describe the pattern of distribution of active volcanoes.

The distribution of active volcanoes

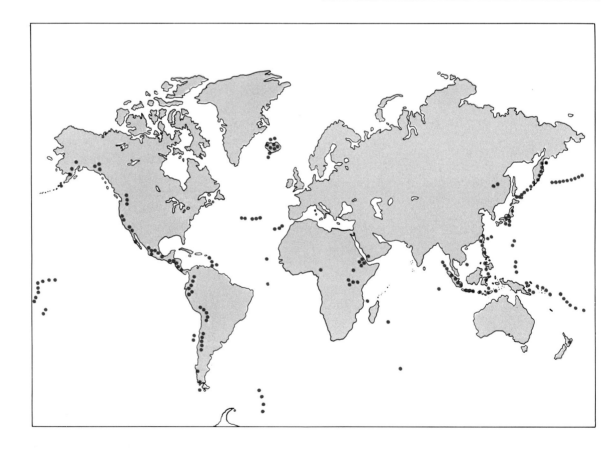

Section 2.3

PUSHES AND PULLS

Folding

Faulting

In normal use, we expect objects such as a drinks can, an iron bar and a plastic jug not to change shape. However, if the conditions of temperature and pressure are changed, these objects can be put under stress. The result of stress is strain, which may bring about a change of shape, or deformation.

F. Observe and record

- Discuss whether heat, pressure or both heat and pressure have been involved in the changes to the deformed objects in the illustration.
- Describe how heat and/or pressure has brought about these changes.
- Draw some other examples of objects that have been changed by heat or pressure, and explain how they have become deformed.

We usually think of the rocks which make up the Earth's crust as being solid and unchanging but if you put rocks under enough stress then they will deform. The type of deformation will depend on the type of force, whether it is compression (a push) or tension (a pull), and the direction of the force. The type of rock is also important. Crystalline igneous rocks such as granite are hard and brittle and may fracture forming a fault. Softer sedimentary rocks may crumple forming a fold.

There are three ways in which a material can deform:

> brittle deformation where a material breaks (fractures) under stress;
>
> plastic deformation where a material changes its shape when a stress is applied – in this case the material stays in the new shape even if the stress is removed;
>
> elastic deformation where the material changes shape when stress is applied but returns to its original shape when the stress is removed.

G. Discuss

- Predict which of these materials would deform in the ways listed above:
  Plasticine;
  rubber;
  chalk.
- Test your predictions.
- List some more materials which would show:
  brittle fracture;
  plastic deformation;
  elastic deformation.

EARTH : **EARTH AND UNIVERSE**

Rocks can behave in a similar way to the materials in your investigation. They can behave in a plastic way and form folds, or they can behave in a brittle way and rupture, forming faults. At the surface of the Earth, where temperatures are relatively low, there are no rocks which behave in an elastic way. The deeper you go beneath the surface the higher the temperature rises, and the behaviour of rock changes allowing for elastic deformation.

If rocks are subject to compressional forces (forces of pushing together) in the Earth's crust they can become folded. A wide range of fold structures can be found but there are two basic types: anticlines (upfolds), and synclines (downfolds).

Clearly the type of the folding depends on the amount of compression and the direction of the forces operating. Forces in the Earth's crust can produce a very small local folding or produce mountain ranges such as the Alps, the Andes, the Himalayas and the Rockies. All the major mountain ranges on Earth were formed by folding.

Folding can occur if the rocks are plastic. Often they are brittle and compressional and tensional forces produce fractures in the Earth's crust. The rocks on one side of the fracture then move relative to those on the other side to produce a fault. Faults can have horizontal or vertical movements.

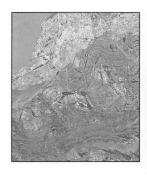

Large scale fold mountains of central Alps

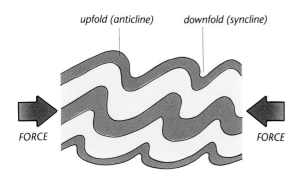

Compression causes folding

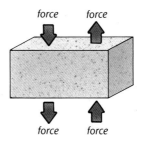

How a fault is produced

During the development of folds the compressional forces can be so intense that the rocks fracture producing a thrust fault. Thrust faults are nearly horizontal. They are associated with the formation of fold mountains where there are intense forces in operation. Along the Moine Thrust in north west Scotland rocks have been moved 10 kilometres from their original position.

Reverse faults are caused by less intense compressional forces than thrust faults. The rocks are pushed together and one block is pushed up over the other.

H. Investigate

- Use blocks of wood to investigate what effect changing the amount and direction of compression has on layers of felt representing layers of sedimentary rock.
- Draw each of your investigations showing clearly the direction of the force.

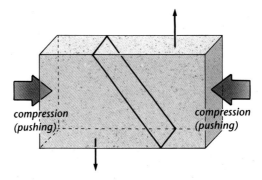

A reverse fault

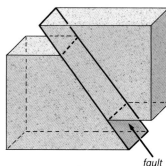

Tensional forces produce the most common type of faults. These are called normal faults. The rocks move apart at the fault and the rocks on one side often move down relative to the other. Normal faults are reponsible for the formation of rift valleys, where a central block has moved down between two normal faults. The East African Rift Valley runs almost the entire length of East Africa.

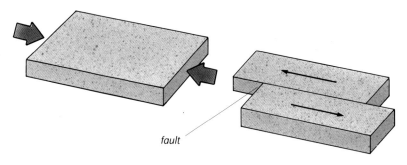

fault

A tear or transform fault

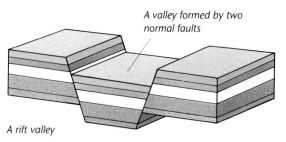

A valley formed by two normal faults

A rift valley

The folded and faulted areas of the Earth's crust today, like those of volcanic actvity, are confined to certain narrow active belts or zones, which surround larger relatively inactive areas of crust.

Normal and reverse faults can have vertical movements of up to hundreds of metres, depending on the forces involved. Horizontal movements along faults can be hundreds of kilometres. Faults with horiontal movements are called tear or transform faults. These occur where blocks of the Earth's crust move past each other. The San Andreas fault in California is an example of such a fault.

I. Record

- Transfer the information from the diagram below to your world map.
- Describe how the areas of deformation relate to the other features on your map, such as volcanoes.

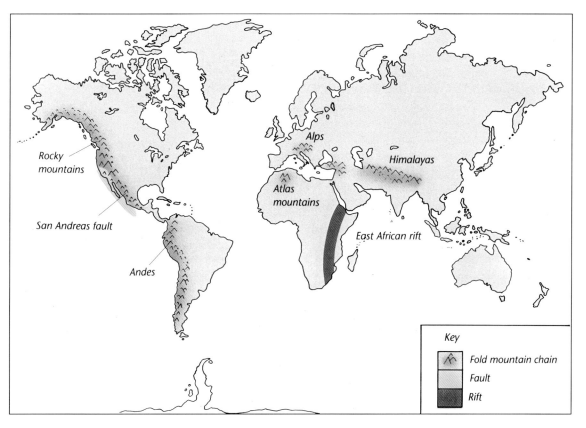

Major structural features of the Earth's surface

EARTH : **EARTH AND UNIVERSE**

Section 2.4

ALL CHANGE

The high temperatures and pressures which can act within the Earth's crust can cause one rock type to change into another. The process that causes rocks to change like this is called metamorphism. The rocks produced are called metamorphic rocks. Although high temperature and pressures are involved the changes happen while the rock remains solid.

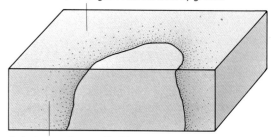

Baked zone around a granite

J. Discuss

1. Where on the Earth would you expect to find metamorphism taking place?

2. What processes are acting in these areas to produce such high temperatures and pressures?

3. Sedimentary rocks are formed at low temperatures and pressures. Igneous rocks are formed at high temperatures. Which of these rock types do you think would be particularly affected by metamorphism?

For the pressure to be high enough for metamorphism to occur there needs to be compression. This occurs during the deformation of rocks to form fold mountain chains. Pressure also increases with depth because of the increasing mass of rock above. Temperature also increases with depth, at a rate of about 100°C for every three kilometres. High temperatures are also caused by the intrusion of magma.

The type of metamorphic rock which is produced depends on the original rock type and the length of time over which metamorphism has to act. There are two types of metamorphism: thermal metamorphism and regional metamorphism. Thermal metamorphism is caused by the high temperatures involved in igneous activity; and is limited to the 'baked' zone around an intrusion; pressure is of only minor importance. The size of the baked zone depends upon the size of the intrusion and the time it takes the magma to cool. Cooling may take millions of years.

K. Observe and record

- Collect the following pairs of original and metamorphic rock specimens:

  limestone and marble;

  sandstone and metaquarzite;

  mudstone and hornfels.

- Compare each pair of rock types. You could look at the following properties:

  hardness;

  porosity;

  density;

  colour;

  the size and shape of grains or crystals;

  the arrangement of grains or crystals.

Those rocks closest to the intrusion show a greater change than those further away. These variations within the baked zone are particularly clear when a magma has intruded into a fine grained rock such as mudstone or shale. These contain clay minerals which have a complex chemistry and therefore have a greater scope for change than say limestone which has a simple chemistry.

L. Work out

- Use Worksheet EU9 *Thermal metamorphism* to show how rocks are changed.

Regional metamorphism involves high pressures. High temperatures can also be important. Regional metamorphism affects large areas and is associated with the formation of fold mountains. When mountains are formed the rocks deep inside them are metamorphosed by the effects of heat and pressure. Much of the Scottish Highlands are made up of metamorphic rocks. These rocks are part of the inside of a mountain chain formed about 1000 million years ago.

As with thermal metamorphism, there are different levels of regional metamorphism. New rocks are formed containing minerals different from those of the original rock. These new rocks have mineral layers or bands. This type of metamorphism can destroy features like bedding and fossils in the original rock.

Under conditions of high pressures but low temperatures rocks rich in clay minerals, such as mudstones, are changed into slate. The clay minerals are altered into new minerals called micas. The tiny flakes of mica are formed at right angles to the directions of pressure. It is this alignment which enables slate to be split apart to make roofing slates.

When high pressures are accompanied by high temperatures the clay minerals in a mudstone change to form large crystals of mica which form parallel bands. The rock type produced is called a schist.

Schist

Gneiss

Deep within a fold mountain there are very high temperatures and pressures. The rock type produced is called a gneiss and shows alternating bands of dark and light minerals.

At these high pressures and temperatures the rocks are near melting point. If melting does occur granite can be produced at the heart of the mountain chain.

The formation of slate

Slate can easily be split

M. Work out

- Use Worksheet EU10 *Regional metamorphism* to show how new rocks are formed.

Section 2.5

MAKING WAVES

Damage caused by the San Francisco earthquake in 1989

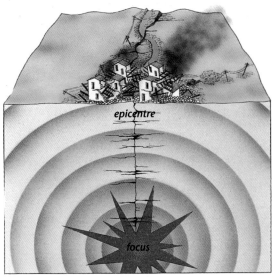

The focus and epicentre of an earthquake

When it is put under enough pressure, a brittle material like rock often breaks. As it ruptures, the force which is being applied is released as energy in the form of shock waves (seismic waves). These seismic waves cause earthquakes. Hundreds of earthquakes happen every day, some are small tremors and others large, full-scale earthquakes. Most occur in the oceans or in uninhabited areas and go unnoticed. Occasionally large earthquakes occur in populated areas of the Earth and cause immense amounts of damage and loss of life. The Kanto earthquake in Japan in 1923 claimed the lives of 142 800 people.

The point at which a fracture occurs under the Earth's surface is called the focus. The point on the Earth's surface directly above the focus is called the epicentre. It is at the epicentre that the shock waves will be felt most strongly. From the epicentre the shock waves travel in all directions, travelling right through the Earth. The magnitude of an earthquake is a measure of the amount of energy which has been released.

Earthquakes produce three kinds of seismic waves. Primary (P) waves, secondary (S) waves and surface waves. P waves are produced by pushing and pulling forces, and cause the rock to shake backwards and forwards. They are longitudinal waves. P waves travel at about six kilometres per second and can travel through solids and liquids. S waves cause rock to shake at right angles to the direction of movement of the wave. They are transverse waves. S waves travel at three kilometres per second and cannot travel through liquids. Both P and S waves can travel through the Earth, and so are referred to as body waves. Surface waves can only travel in the crust of the Earth. Surface waves have a long wavelength and cause the largest movements and the most damage; they travel more slowly than P and S waves.

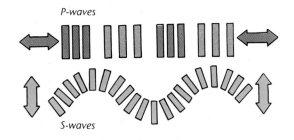

P and S waves produced by a slinky

Earthquakes are detected by instruments called seismometers. These are linked to recording equipment which produce a permanent record of seismic activity. This record is called a seismogram and will show the arrival of the P, S and surface waves.

Seismometer and seismogram

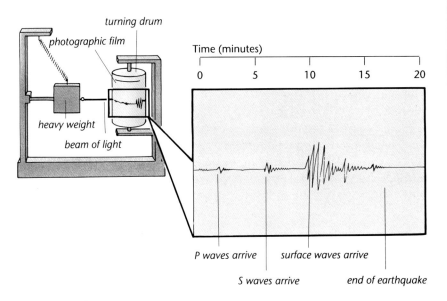

P waves arrive
S waves arrive
surface waves arrive
end of earthquake

N. Work out

- Use Worksheet EU11 *Seismic waves* to find the centre of an earthquake.

The data gathered from the seismic stations around the world has provided information about where, and at what depth, earthquakes occur. Most earthquakes occur at depths of less than 30 km and none are known from below 700 km. As temperature increases with depth so the properties of rock change; they become more 'plastic'. Below 700 km rocks are 'plastic' in their behaviour and therefore bend rather than break to generate earthquakes.

In addition to being confined in depth, earthquakes are generally confined to narrow zones, called seismic zones.

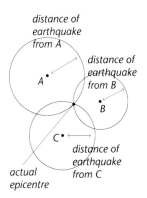

Locating an earthquake with three seismic stations

Since P waves travel faster than S waves they arrive at a seismometer before the S waves. This time difference can be used to calculate how far away the recording station is from the epicentre of the earthquake. Every minute of time difference represents about 1000 km. The greater the time difference the further away the recording station is from the epicentre. This method provides information about distance, but gives no indication about direction. You need information from at least three different recording stations to be able to identify the position of the epicentre.

O. Observe and record

- Plot the distribution of earthquakes on your map of the world.
- What patterns can you see in the relationship between these seismic zones and other features on your map.

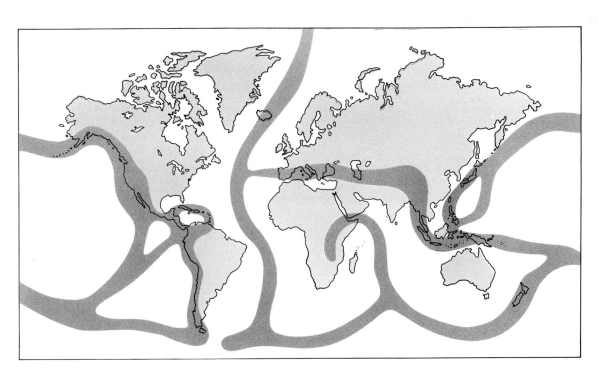

Earthquake zones of the Earth's surface

Section 2.6

INSIDE THE EARTH

We know a great deal about the surface of our Earth. We can collect specimens, observe many of the processes which shape the land surface and make calculations about its size, mass and density. The average density of the Earth is 5.5 g/cm³, but surface rocks only have a density of between 2.5 g/cm³ and 3 g/cm³. There must, therefore, be rocks of greater density inside the Earth.

We cannot directly observe what goes on deep within the Earth but, using evidence particulary from seismic waves, it has been possible to develop a good understanding of the Earth's internal structure.

A seismic survey being carried out in Guatemala

Seismic waves can be used to predict the internal structure of the Earth in a similar way to how ultrasound is used to check on an unborn baby's development.

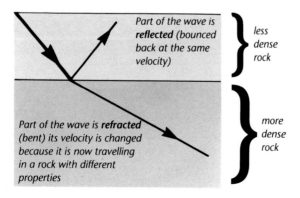

Reflection and refraction of waves

As seismic waves travel through the Earth they pass through layers of rock with different physical properties, such as different densities. The change from one rock type to another may be gradual or there may a distinct change between the layers. The line of change is called a boundary. When a seismic wave meets a boundary some of the wave is reflected at the same velocity, and some is refracted and its velocity changes. By studying these changes in seismic wave velocity seisomologists have shown that the Earth has a layered structure.

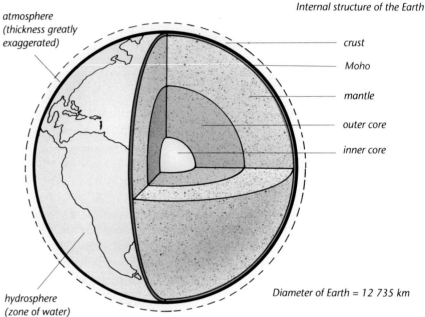

Internal structure of the Earth

Two distinct boundaries, or discontinuities, have been identified marking sudden changes in the properties of rocks. The Mohorovicic discontinuity (or Moho) varies from 6 km to 70 km below the Earth's surface. The layer of rocks above the Moho is called the crust and the layer below the Moho is called the mantle.

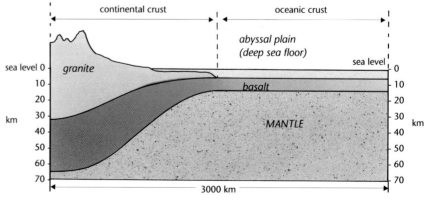

The Earth's crust

Seismic data shows that the crust can be divided into two types; continental crust which forms the earth's continents and oceanic crust found beneath the deeper parts of the oceans. Continental crust contains a wide range of rock types which date back to over 3500 million years. The overall chemical composition is similar to that of granite. It has an average thickness of about 30 km, but under developing mountain chains such as the Himalayas is as much as 70 km. The average density for continental crust is 2.7 g/cm^3.

Oceanic crust is on average only 6 km thick. Its chemical composition is similar to that of basalt, and it has a density of about 3.0 g/cm^3. No oceanic crust is more than about 200 million years old.

The second discontinuity occurs at a depth of 2900 km. This is the core-mantle boundary. The composition of the upper part of the mantle is known from blocks of rock carried to the surface by volcanic action and, more rarely, where mountain building processes and then erosion have exposed small parts of the mantle. The rock type found is called peridotite, and has a density of about 3.5 gram/cm^3. The composition of the lower part of the mantle is not known but is thought to be a very dense type of peridotite.

The behaviour of P and S waves at the core-mantle boundary provides information about the size and composition of the core. In the next activity light is used to model the effect the core has on P waves.

P. Investigate

- Draw a circle 12 cm in diameter onto a piece of black paper.
- In the centre of the circle place a 250 ml beaker containing a rubber bung and partly filled with water.
- Place the ray box at what would be the North Pole.
- Observe carefully the pattern of light around and inside the beaker. Notice how the light is bent (refracted) as it passes into and out of the water.
- Using a piece of chalk, mark those areas of the circle which receive light and those which are in shadow.
- Draw and describe the pattern you have observed.
- Now use the slits. Adjust the position of the ray box so the two outer rays just touch the beaker. Draw and describe what you see.
- Experiment by using beakers of different diameters and with and without water.

Peridotite

A large earthquake which had its focus at the North Pole would be recorded normally as P, S and surface waves in seismic stations all over the northern hemisphere. In the southern hemisphere, only P and surface waves would be recorded in Antarctica, and only surface waves in New Zealand, Australia, South Africa, and the southern part of South America. There is a zone, called the shadow zone, where P and S waves are not received.

Information from sesimic study suggests that the core has an outer liquid part and an inner solid part. The outer liquid part is thought to consist of molten iron and iron sulphide with a density of about 10 g/cm^3. The inner solid part is thought to be made up of an iron-nickel alloy with a density of about 13.5 g/cm^3.

Two other lines of evidence provide clues to the composition of the inner parts of the Earth. The fact that the Earth has a magnetic field suggests that the core of the Earth is iron rich. Meteorites which occassionally fall on Earth are thought to represent bits of planetary material which formed at the same time as the Earth. Each of the three main types of meteorite has a composition and density similar to that suggested for the different layers of the Earth; there are iron-nickel alloy meteorites, and those with similar compositions to the mantle and the crust.

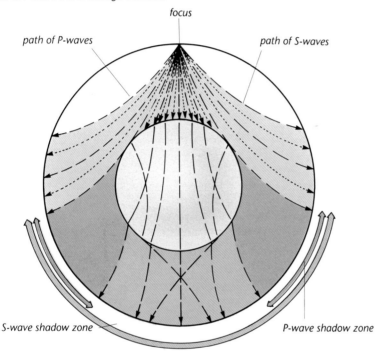

The paths of P and S waves through the Earth

R. Work out

- Use Worksheet EU12 *Inside the Earth* to write a summary of the Earth's structure using the information in the text.

In addition to providing information about the internal structure of the Earth the study of seismic waves has played a part in the development of a theory which has revolutionised the study of the Earth. This is the theory of plate tectonics. The next unit shows how this theory was developed. It uses all the information you have gathered about different rock types and how rocks can be deformed.

Q. Think about

1. How does the size of the shadow zones give an indication of the diameter of the Earth's core?

2. What can you tell about the state of the core from the fact that S waves are not received in a large part of the southern hemisphere.

UNIT 3
FOUNDATIONS

Section 3.1

DRIFTING CONTINENTS

As early as 1620 Francis Bacon, one of the first experimental scientists, noted that the shapes of the coastlines of Africa and South America were very similar.

Francis Bacon

edge of continental shelf

The fit of South America and Africa

A. Discuss

- Why do you think there is a better fit between the continental shelves of South America and Africa rather than their coast lines?

During the seventeenth and eighteenth centuries there were a number of suggestions made for the shape and distribution of the oceans and continents. Some took the biblical view that the oceans had been formed at the time of Noah's flood. Others suggested that the Earth was expanding, or that there was major faulting. There was even a suggestion that the Pacific ocean was formed when the moon was torn away from the Earth.

In 1915, a German meteorologist called Alfred Wegener published a book on the *Origin of Continents and Oceans*. In this book he presented the idea of continental drift and used evidence from palaeontology (the study of fossils), geology (the study of rocks) and meteorology (the study of climate) to support his hypothesis.

Similarities in the geology of South America and Africa

It is not just the coastlines of South America and Africa which match. There is also a similarity in the geology. The rocks and the structure of the two land masses is the same. There is match between ancient areas and old mountain belts.

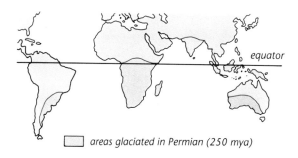

Areas glaciated 250 million years ago

EARTH : **EARTH AND UNIVERSE** **85**

There is evidence that Antarctica, South America, Southern Africa, Australia and India were once partially covered by the same ice sheet. There are thick deposits of glacial sediments and evidence of glacial erosion all of the same age, Permian (250 million years old). There are also large boulders (called erratics) found in Brazil which are of rock types only found in Africa. These large boulders could have been carried to South America by the ice sheet and deposited there as the ice sheet melted and retreated.

B. Think about

1 Explain how it would be possible to have glaciation in different areas of the world at the same time if:

 a the continents have been moved apart;

 b the continents have always been in their present position.

2 Which of these hypotheses is the easiest to explain?

3 How could these hypotheses be tested?

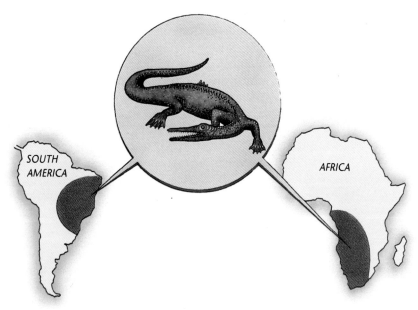

Mesosaurus in South America and in Africa

The distribution of fossil plants and animals provide more evidence which supports the idea of moving continents. The study of fossils has shown that similar plants and animals lived on the southern continents which are now separated by thousands of miles of oceans. *Mesosaurus*, an alligator-like reptile, is found in lake sediments, 270 million years old, in both South America and Africa. Since it lived in fresh water it is unlikely that *Mesosaurus* could have migrated across a wide ocean, suggesting that Africa and South America were once joined. Similarly unlikely is the spread across oceans of a fossil fern called *Glossopteris*, which is found in rocks 200 million years old on all the southern continents including India.

Living plants and animals also provide evidence which supports continental drift. Darwin's theory of evolution tells us that plants and animals which are similar must have evolved from a common ancestor. Marsupials (mammals with pouches, such as the kangeroo) are only found today in America and Australia. It is unlikely that marsupial mammals evolved in two separate places. This evidence suggests that South America and Australia were once joined.

There is similar evidence that supports the idea that continental masses of the Northern Hemisphere were also joined in the past.

C. Record

- Use Worksheet EU1 3 *Evidence for continental drift* to make a summary of the information from geology, glaciation and fossils which you have read about in this section.

Wegener presented his ideas at an international meeting in New York in 1926. Wegener suggested that the less dense continental rock 'floated' over the more dense oceanic rocks, and that this movement was powered by gravitational forces. He even suggested that during one period of Earth history all the continents were joined in a super continent called Pangea. There was some interest expressed in his ideas but there was a general lack of support within the scientific community for the idea of continental drift. It was not until the 1950s, long after Wegener had died, when new evidence was provided from the ocean depths, that his ideas were taken seriously.

Section 3.2

SEA FLOOR SPREADING

During the Second World War submarines played an important role in the conflicts. Just as you need information about the Earth's surface to help you to navigate on land, so it is necessary for submariners to have a map of the sea floor in order to navigate safely. The technology was developed to make underwater maps using echo sounding (sonar). Instruments that detected changes in magnetic field were developed to help submarines locate other vessels magnetically.

This technology was later used to study the oceans in detail. By the mid-1950s detailed maps of the ocean floors were produced. The diagram shows the major features found all over the Earth's oceans.

> **D. Work out**
>
> - Write out descriptions of the ocean floor features labelled in the diagram on your glossary worksheet, EU32.

A system of ocean ridges totalling some 80 000 km in length has been identified. All of the world's oceans have an ocean ridge system. These ridges have a central rift valley. This valley is produced by faulting, giving high sides and a lower central section. There is a great deal of volcanic and earthquake activity associated with these ocean ridges. They do not form a continuous mountain chain. The ridge is offset by a series of faults called transform faults.

Occasionally an ocean ridge rises above sea level. An example is Iceland, which lies on the North Atlantic Ocean ridge. A rift runs along the length of Iceland and is associated with a lot of volcanic activity and earthquakes.

Major features of the ocean

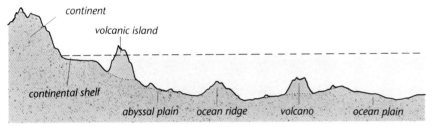

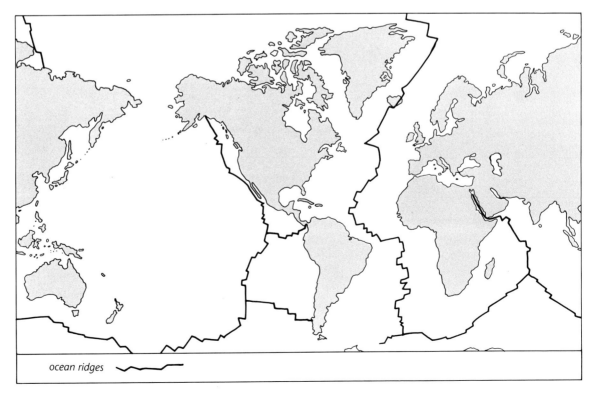

Distribution of ocean ridges

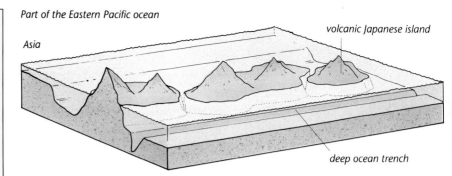

Part of the Eastern Pacific ocean

Asia

volcanic Japanese island

deep ocean trench

E. Record

- Transfer the information from the map on the previous page onto your world map on Worksheet EU8.
- Describe the pattern of distribution of ocean ridges.

  How does this relate to other features on your map?

Japan is made up of a series of islands. These islands are of volcanic origin and are subject to frequent earthquakes. Such chains of islands are called island arcs. On the ocean side of these island arcs are found the ocean trenches. These trenches can be up to 11 km deep and run parallel to the island arcs.

The Pacific Ocean, in addition to having an ocean ridge system, contains two other important oceanic features – island arcs and ocean trenches.

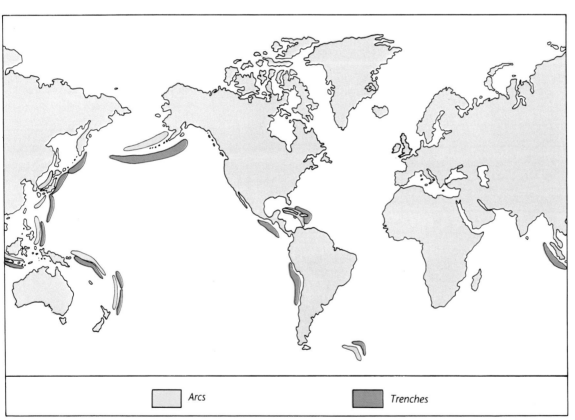

The distribution of island arcs and ocean trenches

Arcs · Trenches

F. Record

- Add the positions of island arcs and ocean trenches to your world map on Worksheet EU8.

  What patterns can you see in their distribution?

A study of the pattern of earthquakes in an island arc/ocean trench system shows that earthquakes are found at progressively greater depths on the continental side of the trench. They are also confined to a rather narrow zone which usually slopes at an angle of 45°. Earthquakes have been identified along the zone to depths of 700 km.

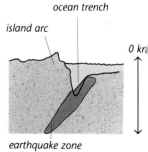

Pattern of earthquakes at an ocean trench

The study of variations in the Earth's magnetic field has also provided evidence for drifting continents and sea-floor spreading.

The Earth's magnetic field shows that the Earth behaves in a similar way to a bar magnet. The axis of the Earth's magnetic poles is at an angle of about 11° to the Earth's axis. The Earth's magnetic field is thought to be produced as a result of electric currents flowing in the outer core.

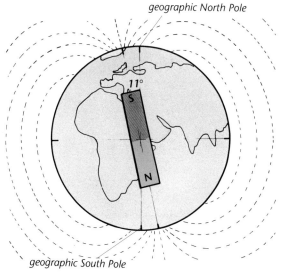

The Earth's magnetic field is shaped like that of a bar magnet

Throughout the 1950s and 1960s detailed magnetic surveys were done of the oceans. Maps showing the pattern of magnetic highs and lows were produced. A magnetic high occurs where rocks show a magnetic effect greater than normal, and a magnetic low where rocks show a magnetic effect less than normal. On such maps high anomalies are shown in black and low anomalies in white.

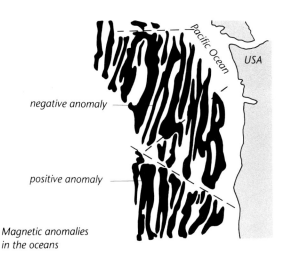

Magnetic anomalies in the oceans

As the result of studying magnetic data from all over the world, two British scientists Fred Vine and Drummond Matthews suggested in 1963 that the magnetic highs and lows were caused by reversals in the polarity of the Earth's magnetic field, i.e. when the north magnetic pole became the south magnetic pole, and visa versa. Highs occur where rocks were magnetised in the same direction as the Earth's magnetic field at that time. Lows resulted when the rocks were magnetised in the opposite direction. Why these magnetic reversals take place is not fully understood.

a
As new rock is formed it takes up orientation of magnetism of that time

b
After a magnetic reversal new rock takes up new orientation of magnetism

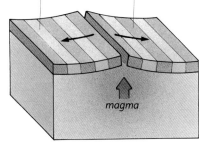
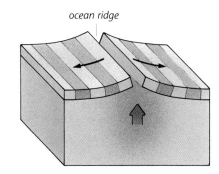

Reversals in a magnetic field

G. Work out

- Study the sand model of an ocean ridge. A number of bar magnets have been buried in the sand to represent magnetised rocks on either side of the ocean ridge.

- Using a plotting compass to represent an instrument for detecting magnetism in rocks, a magnetometer, work out the pattern of changes.

- Draw your own map of magnetic changes.

- Describe the pattern. (There are two important things you should notice.)

EARTH : **EARTH AND UNIVERSE**

As well as being able to see a pattern in the rock magnetism, it is also possible to find out in what latitude rocks were when they were formed. This only works with igneous rocks, particularly those like basalt which contain iron minerals. As a lava cools the iron rich minerals line up with the magnetic field in which they are formed. When the lava becomes solid this alignment is fixed into the rock, and can be measured.

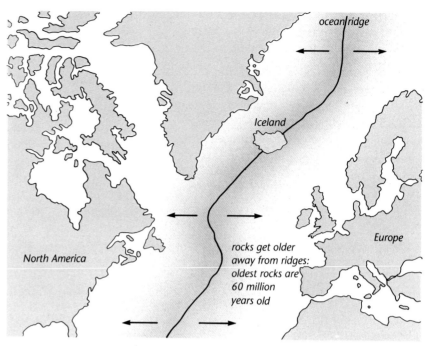

Ages of rocks under the Atlantic

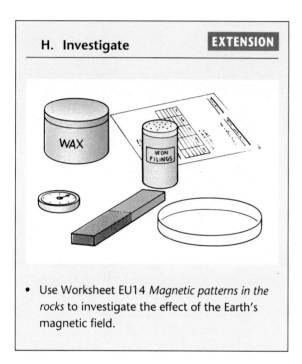

- Use Worksheet EU14 *Magnetic patterns in the rocks* to investigate the effect of the Earth's magnetic field.

The study of the magnetic orientation of rocks of different ages on the same continent showed that these rocks gave different positions for the magnetic pole. The pole appears to have changed position through time. However it is not the poles that have changed position but the continents that have moved relative to the poles.

The technology developed for deep-sea drilling for oil and gas was put to use in the late 1960s. In 1968 the scientists on board the drilling ship the *Glomar Challenger* began collecting rock samples from the deep oceans. These rocks were identified and dated using radiometric techniques. The data collected in this way showed two things: the oceans are relatively young, with no oceanic rocks being older than 200 million years, and the oldest oceanic rocks found get progressively older away from ocean ridges.

By the 1970s scientists had developed the ideas of the previous forty years on continental drift and sea floor spread into the theory of plate tectonics. This theory states that the Earth's crust is made up of a number of rigid plates, and that these plates are moved relative to one another. The margins, where the plates are being moved in different directions, are the geologically active zones.

I. Think about

- Working in a group use the maps and other information you have gathered from this unit and the previous unit to prepare a presentation describing the evidence which suggests that:

  the crust is made up of a number of plates with active margins;

  these plates are moved relative to one another.

Section 3.3

PLATE TECTONICS

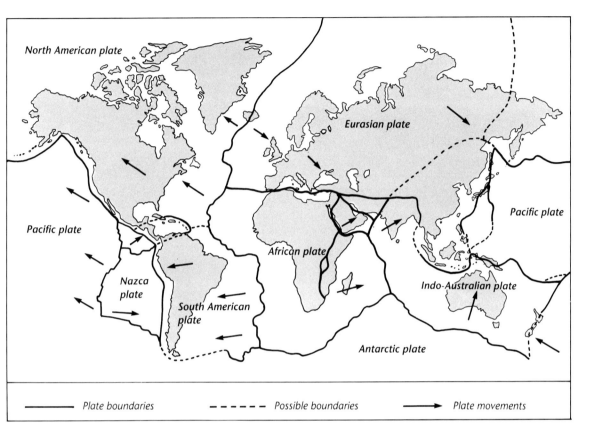

The major plates of the crust

The outer part of the Earth is formed of plates. Each plate is about 100 km thick, and is made up of the crust and part of the upper mantle. This is known as the lithosphere. Beneath the lithosphere there is a narrow layer called the asthenosphere. This part of the upper mantle behaves in a plastic way and movement of it causes the movement of the plates.

Mountains, trenches and ocean ridges mark the boundaries of the plates.

There are three types of plate margin recognised:

constructive margins where new crust is being created;

destructive margins where crust is being destroyed or modified;

conservative margins where plates are sliding past each other.

Each of these different types of margin produces distinctive geological features.

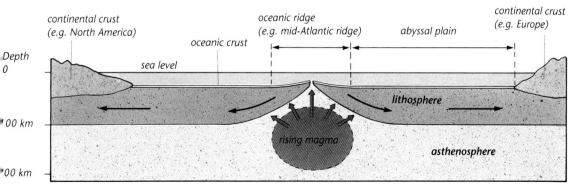

Constructive plate margin

EARTH : **EARTH AND UNIVERSE** **91**

The 80 000 km of ocean ridges around the world mark constructive plate margins. At constructive margins magma from the asthenosphere rises us into the ridge system where it forms basalt lava flows. These basalts form the upper parts of the new ocean crust. Magma which cools below the surface forms the lower parts of the new oceanic crust, and consists mainly of the coarse-grained rock gabbro.

As new oceanic crust is produced and added to the plates so it is forced away by the next batch of new crust. In this way the sea-floor is spread and the continents on either side of the ocean move further apart. The Atlantic Ocean is spreading at a rate of 2 cm/year. Constructive margins are associated with shallow earthquakes to depths of 14 km.

Destructive plate margins can usually be recognised by the presence of an ocean trench. These mark places where two plates are moving towards each other. The edge of one of the plates is forced down beneath the edge of the other, this process is called subduction and produces a subduction zone.

There are three possible types of destructive plate margin:

   those which occur when two plates carrying oceanic crust meet;

   those where a plate carrying a continent meets one carrying oceanic crust;

   those between two plates carrying continental crust. (Originally there would have been oceanic crust between the two plates but this has been subducted and lost.)

Where two plates carrying oceanic crust meet, an island arc/ocean trench system is produced. As the descending subducted plate is forced down, the combination of increasing depth and friction causes melting. The basaltic upper part of the oceanic crust starts to melt and be absorbed at depths of 100-300 km. Melting and absorption of all the lithosphere is completed by 700 km. The less dense parts of the magma rising from the subduction zone reaches the surface to produce a chain of volcanic islands. Japan is an example of this process.

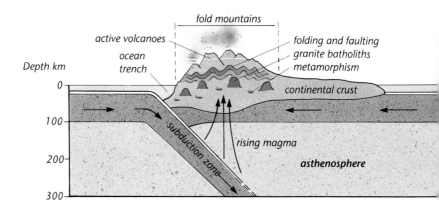

An ocean/continent margin

In the situation where a plate with oceanic crust meets a continental margin, it is always the oceanic plate which is subducted. Oceanic rocks are denser than continental rocks. During this type of activity two things happen; fold mountains are produced and there is much volcanic activity. The fold mountains are formed from the marine sediments deposited on the continental shelf and slope. These sediments, which can be up to 5 km thick, are trapped between oceanic plate and the continent. This compression causes the sediments to deform and fold. They rise up above sea level forming a fold mountain chain.

The Andes mountains in South America were formed in this way. Fossils of marine animals can be found on the highest peaks. The intense compressional forces can produce intense regional metamorphism, producing granites, gneiss and schists. Folding is also intense, and thrust faults can occur. The volcanic activity results from the melting oceanic plate. The most common lava produced is andesite named after the Andes. Eruptions of rhyolite also occur. Magmas continue to rise and form intrusions with localised thermal metamorphism.

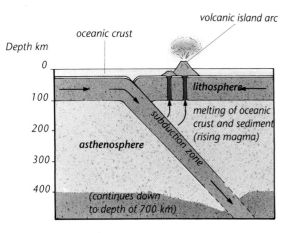

An ocean/ocean plate margin

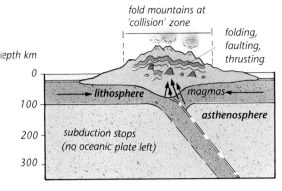

A continent/continent plate margin

The most well known example of the result of a collision between plates carrying two continents is the Himalayan mountain chain. About 200 million years ago, what we now call India was joined to the west coast of Africa. From this time India has moved gradually northwards as the oceanic crust between it and Asia was subducted. Eventually all the oceanic crust was subducted and the continents of India and Asia collided producing the Himalayas. This processes is still continuing; Mount Everest climbers have a centimetre further to go each year!

The Himalayan mountain chain, like the Andes, is made up mostly of marine sedimentary rocks which have been uplifted and intensely folded and faulted. These sediments would have originally been deposited in the ocean between India and Asia. Regional metamorphism occurred producing granites, gneiss and schists. There was also intense volcanic activity, the result of magma produced by the melting of the descending oceanic crust. Although subduction eventually stops, igneous activity continues for some time as the remaining oceanic crust is melted and absorbed into the mantle.

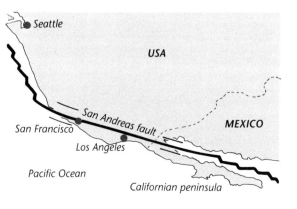

A conservative plate margin

If you lived in California you would be well aware of what a conservative plate margin was, because you would be living on top of one. The San Andreas fault which runs through California marks the point where the Pacific plate and the North American plate are sliding past one another. Such a fault is called a transform fault. No crust is being created or destroyed but the plates are moving past each other at a rate of 5 cm/year.

The intense frictional forces between the two plates means that this area is subject to numerous earthquakes. During the 1906 earthquake the fault moved four metres and produced a destructive earthquake. In trying to predict when earthquakes will occur data on statistics and probability are used. Some areas along the fault have probabilities of a major earthquake occurring in the next 30 years of up to 60%.

Aerial view of the San Andreas fault

J. Work out

- Read the text in this section carefully and write a summary under the following headings (an example has been done for you):

Type of plate margin	Examples	Features Produced	Earthquakes and deformation	Igneous activity	Metamorphism
conservative	San Andreas fault	changes in position	earthquakes	no	no

The most widely accepted explanation of the mechanism for plate movement is called the convection current theory. The diagram shows the convection currents set up in a pan of water when is is heated. As the water is heated its temperature rises and the water moves upwards being replaced by colder water. As the water rises it cools and sinks. This circular motion is called a convection cell.

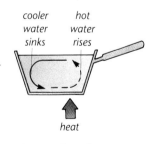

A convection cell

K. Observe

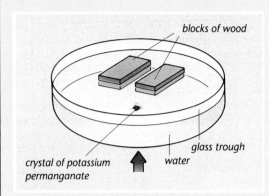

- Watch a demonstration of the effects of convection. The blocks of wood represent continents, and the water represents the mantle.
- As the water is heated carefully observe the dye produced from the crystal and what happens to the blocks of wood.
- Draw diagrams of the two movements, labelling the convection currents produced, and the movement of the 'continents'.
- Use Worksheet EU15 *Convection currents* to draw in the pattern of convection cells that may exist in the mantle.

The convection current theory goes a long way to explaining plate movements. The loss of crust at subduction zones being balanced up by the production of new crust at ocean ridges. There is still a great deal to be discovered about the mechanism of plate movements, as you have seen the pattern of subduction zones and ocean ridges is not a regular one.

Earthquake evidence suggests that much of the mantle may be too rigid to allow convection currents, and that these may be confined to the asthenosphere.

Gravity also has a role in plate movements. Oceanic plates are pulled down under the influence of gravity.

Section 3.5

EARTH HISTORY AND FUTURE

The fact that the Earth's crust is made up of a number of plates which are moving, has not only influenced the pattern of the geology, including the distribution of resources, but also evolution of life and the pattern of the Earth's climate.

L. Present

- Use Worksheet EU16 *The Moving Earth Show* to present a picture of how, throughout the history of the Earth, the continents have gone through periods of joining to form larger 'super continents' and periods of break up into smaller continents.

When the continents are grouped together the climate shows less seasonal change than when they are split up. During periods of rapid sea-floor spreading sea level rises. This is because oceanic ridges occupy a lot of space and so sea level rises to accommodate them.

When the continents are joined into one land mass there is much less continental shelf than when they are split into several smaller land masses. As a result there is much greater competition between marine plants and animals, a reduction in the diversity of the flora and fauna, and the extinction of many species. By the end of the Permian era, some 230 million years ago, up to 70% of all marine species became extinct. Similarly the reason why only the pouched mammals (marsupials) are native to Australia, (rabbits and other non-pouched mammals were introduced by man), is that Australia had moved away from other continents before the non-pouched mammals had the opportunity to spread to Australia.

In the British Isles the presence of resources such as coal, oil, gas, metals and salt are directly link to plate tectonics. The oldest rocks in Britain are 2900 million years old, and the youngest are still being formed. During this long history Britain has been associated at various times with active plate margins. It has

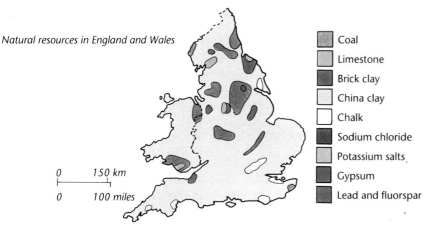

Natural resources in England and Wales

- Coal
- Limestone
- Brick clay
- China clay
- Chalk
- Sodium chloride
- Potassium salts
- Gypsum
- Lead and fluorspar

0 — 150 km
0 — 100 miles

M. Discuss

- Why does Britain, a relatively small country, have such a wide range of geological resources?

---

been subjected to dramatic changes in sea level, and has drifted through different climatic belts. The present shape of the British Isles has only existed for about 10 000 years.

During periods of association with active plate margin, igneous activity led to the formation of metal ore deposits. The granites of northern and south western Britain, for example, are associated with deposits of copper, zinc, lead and tin. These used to be important economically in the past but are no longer worth mining.

Throughout its history Britain has been moving steadily northwards from a position 40° south of the equator to its present position 55° north of the equator. This has had a big effect on the landscape, flora and fauna and the sedimentary rocks formed. During the later part of the Carboniferous period, about 300 million years ago, much of Britain lay on the equator and was covered by densely vegetated swamp. It is this plant material which has been buried to produce our coal deposits, and which is the source of gas in the southern North Sea.

About 250 million years ago Britain was in a position similar to that of the Sahara. Much of it was covered by desert, and later by a shallow sea. This sea was prone to evaporation and thick deposits of salt developed. This salt includes gypsum and rock salt, both of which are important in the chemical industry.

For much of Britain's more recent history it has been covered by seas which have formed limestones such as the chalk of southern Britain. Limestone has a wide range of important uses in the chemical industry and in construction. The organic rich sediments produced in seas about 150 million years ago are the source of North Sea oil.

The relationship between resources and plate tectonics is not confined to Britain. The pattern of distribution of resources is closely linked with the plate tectonic history of any area. The copper belts of the USA and of South America are related to subduction zones which occurred in these areas 150-200 million years ago. The Red Sea which is part of a relatively new constructive plate margin shows deposits of iron, zinc, copper and lead.

The movement of plates will continue for hundreds of millions of years. It is difficult to predict what long term changes these plate movements may have on the evolution of life on Earth. It is possible to predict what pattern the Earths continents will have in the future. This is possible because we can use the direction and rate of movement of plates at present.

N. Work out

- Use Worksheet EU17 *Atlas of the future?* to predict the future shape of the Earth's surface. The following questions will help you think about this piece of work:

  1 Where will new mountain chains be formed?

  2 What might happen to the East African Rift System?

  3 Where might new conservative or destructive plate margins form?

  4 Which parts of the world might be subject to volcanic and earthquake activity?

  5 What might happen to the sea level in the future?

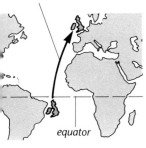

500 million years

The movement northward of Britain through different climatic belts

EARTH : **EARTH AND UNIVERSE** **95**

UNIT 4
JOURNEY INTO SPACE

Section 4.1

LIGHTS IN THE SKY

From the earliest times people have looked up at the sky. Many have seen patterns in the 'heavenly bodies'. The stars seem to be arranged in patterns – the constellations. Regularities in time have been noticed in the phases of the moon and the movement of the stars. These patterns, and also the movement of the Sun, have helped people to navigate and to divide time into days, months and years. Other people, perhaps more scientifically minded, have tried to explain these patterns in terms of models which show how the stars and planets move.

Most people, at some time, have wondered 'where did we come from?' and 'why are we here?'. Many seek answers in the theories of astronomy. The next three units are concerned with the extent to which astronomy can answer questions like these, with observation and pattern making in astronomy, and with models and explanatory theories.

Most of what is known about the universe has been found out by looking into space from the Earth's surface. Observing space needs some skills if you want to see anything that you can make sense of. First consider what some of the problems are.

A. Discuss

- Why can only the Sun be seen during the daytime, and not the stars, moon or planets?
- Why is it hard to see stars even on a clear night when you are in a town or city?
- What differences might you see between planets and stars?
- How do you find out where to look for stars and planets, even if you have a clear, dark sky?

The answer to the last question in activity A is that you can use a star map. Star maps usually contain so many stars that they are confusing. You also need to know your directions.

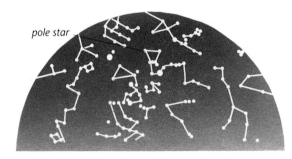

The night sky, looking north in mid-January

The illustration shows some of the brightest stars you might see if you looked north at about 8.30 pm in the middle of January.

Because of the movement of the Earth, this arrangement is always changing. It changes daily because of the Earth's rotation, and it changes annually because of the Earth's orbit. Only one star does not change its position. This star is called the pole star.

One of the greatest observatories of the time before telescopes, erected at Delhi in India. It contained instruments which were capable of yielding very valuable results

B. Observe

- Choose a clear, moonless night away from street lights to observe the sky. Look for different kinds of objects. Aeroplanes are easy to spot!

  Can you see satellites, meteorites or shooting stars and planets?

- Look for the pole star. Compare the stars around the pole star with those shown in the illustration on page 96. Record the position of these stars.

  Can you see any pattern in the arrangements?

- If possible repeat your observations at different times and on different nights. See if the pattern changes. Worksheets EU18 *Observing the night sky* and EU19 *The night sky throughout the year* will help you with these observations.

Herschel's workshop in the museum at 19 New King Street, Bath

C. Discuss

1. In your observations of the moon, planets or stars have you used a telescope or binoculars? If so, how did this improve your observations?

2. Think of some science investigation you have done which required much careful observation. How much did your success depend on the factors listed for William Herschel's work:

   a good planning;

   b the best apparatus;

   c a hard-working assistant – not necessarily your sister?

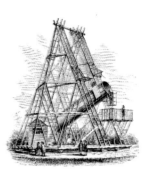

This is a replica of the telescope with which Herschel discovered Uranus in 1781

In astronomy, 'observing' means receiving and interpreting light from planets, stars, galaxies and other objects in space. Often instruments are used to help. The introduction of telescopes in the seventeenth century greatly increased mankind's knowledge of these objects. Galileo was the first to use a telescope for astronomy. You can find out more about his important discoveries in Section 5.1.

For many years astronomers had to make their own instruments (with the help of their servants). William Herschel discovered the planet Uranus with his own telescope, from his back garden in Bath on 13 March 1781. His discovery of this planet, more than two billion miles from the Sun, doubled the size of the universe that was known at that time. He has been called the greatest observational astronomer. Much of his success was due to three things:

- his systematic approach – he mapped and catalogued thousands of stars;

- he made and used the best telescopes of his day;

- his sister Caroline's constant help – she was also an accomplished astronomer, although in those days women did not usually get credit for their work.

A telescope aids observation of the heavens because it makes objects appear larger. By collecting light and focussing it to make an image, it also makes visible objects which could not be seen without it. For a telescope to act as a magnifier, lenses or mirrors of the right strength must be used. These lenses and mirrors need to be as large as possible in order to collect the maximum amount of light. William Herschel realised that it was easier to make very large mirrors than large lenses. He spent many years designing and making these with the help of his brother Alexander and sister Caroline.

D. Work out **EXTENSION**

- Use Worksheet EU20 *The astronomical telescope* to make a telescope and work out its magnifying power.

> ### E. Present   EXTENSION
>
> - Use Worksheet EU21 *William and Caroline Herschel* to find out more about the Herschels' design and use of telescopes.

The world's first radiotelescope, Jordell Bank in Cheshire

The Herschels made an important discovery about radiation during their observations with telescopes. Their investigations showed that stars give off what they called 'invisible light' which has a heating effect. We now call this infrared radiation because it is beyond the red end of the visible spectrum.

> ### F. Investigate
>
> - Devise an experiment to show that infrared radiation is given off from light sources such as the Sun or a filament lamp.
>
>   ⚠ **Do not look directly at the Sun.**

Telescope domes at the 4200 m summit of the Mauna Kea volcano in Hawaii

Astronomers now know that there are many other invisible radiations from space. For example, important discoveries have been made with radiotelescopes and X-ray receivers. These radiations can tell us more about the nature of the universe as we shall see in Unit 6.

For some time telescopes have been sited on high mountains as this improves the view of the sky. More recently it has been possible to view space from outside the Earth's atmosphere completely. In 1957 the USSR launched the first space probe to orbit the Earth. In 1964 an American probe sent back close range photographs of the moon. Since that time probes have been sent to explore all parts of the solar system. They have gathered immense amounts of information and even actual samples of material such as moon rock.

A sample of moon rock

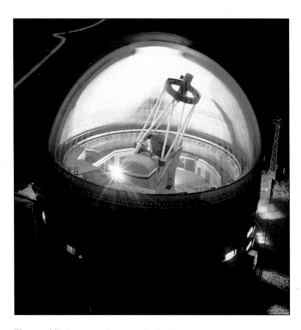

The world's largest telescope, in the Caucasus

> ### G. Think about
>
> 1 Why do you get a much better view of the night sky in the country than in a town?
>
> 2 What is the advantage of making observations from a high mountain or, better still, from a space craft orbiting the Earth? What are the disadvantages?
>
> 3 X-rays and radio waves are less affected by the factors which interfere with our observations – try to find out why.

Section 4.2

INTO THE SOLAR SYSTEM

The first space traveller, major Yuri Gagarin, made his pioneer flight on April 12, 1961 when he orbited the Earth in the vehicle Vostok

Armstrong and Aldrin set foot on the moon for the first time in 1969. This photograph shows Aldrin saluting the American flag

Since the first manned space flight in 1961, over 200 people have been in space. Learning more about the solar system and the universe has not been the only reason for all this activity.

In the past, scientists have often accompanied expeditions run by others. Such expeditions might be motivated by political or military reasons, or by the quest for fame, fortune and adventure. For example Charles Darwin accompanied an expedition organised by the British Admiralty to map the coast of South America between 1831 and 1836. The plants and animals that Darwin saw eventually led him to his theory of evolution through natural selection, published in 1858. The same has been true for the exploration of space. Initial expeditions were known as a 'race' between USA and USSR teams, since at that time the two nations were great political rivals.

Great sums of money were spent on the 'space race' in the 1960s and 1970s. This investment led to a remarkable series of achievements by both nations, and we cannot say which nation 'won the race'. What is clear in the 1990s is that these two nations are now more likely to co-operate. The world political situation has changed, and also neither wants to spend so much money. The next big projects may include a permanent space station in orbit around the Earth, and visits to other planets.

H. Plan and present

Your group is to present a proposal for exploring one of the planets in the solar system.

- Discuss why your mission is important. It will be expensive, so you must have some good reasons. These need not be only scientific.
- Decide where you want to visit. Information about the planets is given in Worksheet EU22 *Planet data* and in reference books.
- Consider the problems of the journey and possible hazards, and decide what jobs different members of your group might do.
- Present your report in writing to the Space Agency (your teacher is a representative and may arrange a selection procedure).

Section 4.3

PLANETARY PATTERNS

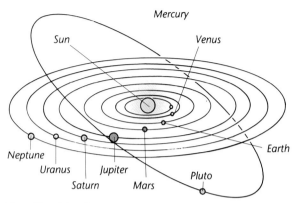

The orbits of the planets

You are now going to look more carefully at the data that is available about the planets. It will help to have the information available as a table, so that you can look for patterns.

I. Work out

- Make a solar system data table showing as many of the facts about the planets as possible, on one sheet. Use the data from Worksheet EU22 and other sources.

  List the planets in order from the Sun: Mercury first, then Venus, ...

  Tabulate the data under headings such as: Distance from the Sun (in millions of km); Diameter (compared to Earth); Mass.

- Look for any patterns in the table.

In 1772 a German astronomer, Johan Bode, publicised a pattern in the distances of the planets from the Sun. Although this had first been proposed by another German, Johann Titius, it is generally known as 'Bode's law'.

J. Work out

- Carry out Titius's calculations using Worksheet EU24 *Bode's law*.

EARTH : **EARTH AND UNIVERSE**

The planets beyond Saturn were not known at the time of Titius and Bode. When Uranus was discovered by Herschel in 1781 it was found to fit the pattern. In fact it was Bode who proposed the name Uranus to Herschel. Astronomers then turned their attention to the curious gap in the pattern between Mars and Jupiter. If the 'law' was to be believed there should be a planet there, about 300 million km from the Sun.

> **K. Think about**
>
> - Use your solar system data table and any other information to predict what the missing planet should be like.
> - You could make up your own data card for this planet (it will need a name!).

In 1800 six astronomers, nicknamed the 'Celestial police', got together to hunt for the missing planet. A new one was discovered. It was only 1000 km in diameter – smaller than the moon – and was named Ceres. It fitted in with Bode's law. Then a second small planet was found slightly further away than Ceres. This was named Pallas. There then followed the discovery of two more planets, which were given the names Juno and Vesta.

> **L. Research**   **EXTENSION**
>
> - Find out what the names Ceres, Pallas, Juno and Vesta mean.
>
>   What do they have in common?

Instead of one large planet, several smaller planets seemed to fill the gap. These are the minor planets or asteroids. It has been estimated that there are about 40 000 of them, which form a belt between Mars and Jupiter. Most are little more than one kilometre across. Occasionally an asteroid passes close to the Earth. In January 1991 the asteroid 1991BA passed within 160 000 km of Earth; astronomically speaking a 'near miss'.

Finding patterns in observations can be a very important step in developing an explanation. Of course there may be more than one possible explanation for a particular pattern. We see the Sun rise in the east in the morning and set in the west in the evening. This pattern can be explained either as the Sun orbiting the Earth daily from east to west (clockwise looking down from the north), or as the Earth rotating daily about a north-south axis (anticlockwise as seen looking down on the North Pole). Perhaps the pattern has no scientific explanation. Although attempts have been made to explain Bode's law, there is no agreement that it is any more than chance that the planets seem to 'obey' it.

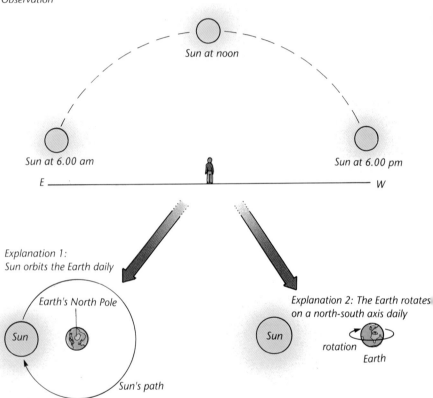

> **M. Interpret**
>
> 1 Why do we now accept the second explanation for our observations of the Sun's apparent movement through the sky? What is wrong with the first explanation?
>
> 2 What possible explanation could there be for Bode's law? Could it be the result of gravity or the way that the solar system formed?

Section 4.4

COMETS AND METEORS

The comet shown in the Bayeaux tapestry is the most famous comet of all. The first known record of it dates back to the Chinese Chronicles of 240 BCE (before common era). For over two thousand years a comet in the sky had been seen as as an omen of impending disaster. It was certainly a bad omen for Harold and the Saxons who lost the battle against the Normans in 1066, as the tapestry depicts. Harold lost his life, and on his birthday too! We now know this comet as Comet Halley, after a famous astronomer who, in 1705, predicted that it would return in 1759.

Bayeaux tapestry, showing Halley's comet

N. Work out

- The dates of some of the more recent sightings of Comet Halley are 1607, 1682, 1759, 1835, 1910, 1986.

  What is the pattern of these dates?

- Predict when Comet Halley should be seen again.

ranging between two and 200 years. At least one new one is identified every year, and they are given the name of the discoverer, often an amateur astronomer. So start looking – you could become famous!

Comet Bennett

There are several theories as to where comets come from. The most widely accepted one is that they originate in the so-called comet cloud. This is beyond the furthest known planet, about 7.5 trillion (7 500 000 000 000) km from the Sun. There are thought to be billions of comets in this cloud, which marks the outer reaches of the solar system. The outermost part of the cloud is a quarter of the way to the nearest star, but the influence of the Sun is still enough to keep the comets in orbit. It is extremely cold so far from the Sun and the comets are like 'dirty snowballs': lumps of ice (frozen gases such as methane and ammonia, as well as water) and dust in perpetual deep freeze.

Occasionally, under the influence of the Sun and other stars in our galaxy, a comet may be pulled out of the cloud and fall towards the Sun. It then takes up a new elliptical orbit that passes much closer to the Sun. About 200 comets have been recognised with periods

Larger comets contain a nucleus which may be rocky, a 'coma' (made of small particles and gas) and the tail of gas and dust.

For most of its orbit a comet is a cold dark body, invisible from the Earth. We can only see it if it gets close enough to the Sun for the solar radiation to cause the comet to glow. Energy from the Sun melts the ice very rapidly, so that gases and dust are lost. On every orbit a comet loses more of its mass and eventually it will vaporise completely.

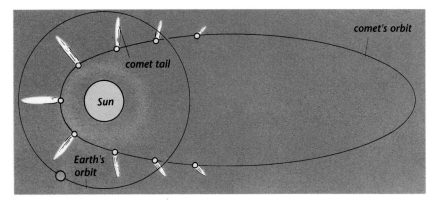

Comet tails as the comet orbits around the Sun

Most of the objects we call meteors are tiny fragments of comets. As the Earth moves in its orbit it may pass through areas where comets once passed and meet the debris. The majority of these small objects are about the size of a snow flake and just as fragile. They burn up when they enter the Earth's atmosphere at speeds of up to 60 km per second, forming 'shooting stars'. Swarms of shooting stars can be seen at certain times during the year. August is a good month to look for them.

Some meteors are larger objects which survive their passage through the Earth's atmosphere and hit the Earth. They are then called meteorites. Most of these come from asteroids. About ten are found each year. Most are quite small by the time they land, but occasionally one is large enough to have quite an impact on the Earth.

Meteorite crater in Arizona

The huge crater in Arizona was caused by a large meteorite that hit the Earth 22 000 million years ago. Some scientists suggest that a huge impact 65 million years ago, perhaps in the Pacific Ocean, caused the major change to life on Earth, including the extinction of the dinosaurs which occurred at that time.

O. Think about

1. Try to imagine what would happen if a large meteorite hit the Earth now. What would be the difference if it came down in an ocean rather than on land? Draw and describe the effects.
2. Many scientists think that the effects of a nuclear war would be similar. It has been called nuclear night or nuclear winter. Why is this?

Section 4.5

BEYOND THE SOLAR SYSTEM

When you look out into space on a clear moonless night you see many stars. Too many even to attempt to count them. Some seem brighter than others. You may start to see patterns in the stars. These patterns are called the constellations. They were first recorded by the ancient Greeks. We know them by Latin names such as Ursa Major, (the great bear), Orion (the hunter), Pegasus, (the flying horse) Leo (the lion) and Taurus (the bull). You will recognise some of these as signs of the zodiac, which astrologers use to predict people's futures. The constellations that make up the zodiac are those that lie in the ecliptic, that part of the sky that the Sun, moon and planets move through. When the Sun is 'in Gemini', for example, it is in the same part of the sky as that constellation.

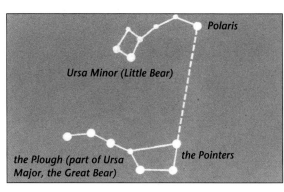

Ursa Major, Ursa Minor and Polaris

For centuries sailors have used the constellations for navigation. Polaris, the pole star, is always directly above the Earth's North Pole. It does not move around like all the other stars – a useful fact if you are lost at night. Other stars and constellations show a regular movement from day to day, and over the course of a year.

A time exposure made with the camera pointing at the pole star; the stars seem to move slowly across the sky, so producing trails. This apparent movement of the stars is due to the real rotation of the Earth on its axis

We now know that the stars we see in the same part of the sky can be at very different distances away from us, and moving at different speeds.

P. Work out

- Make about one hundred dots at random on an A5 size sheet of paper. Go over the dots, making some larger than others.
- Swap your piece of paper with someone else. See if you can find patterns in the dots.
- Join up the dots to make your own constellations. Give these constellations names.
- Move the paper so that one of the dots stays in a fixed position while all the others move. What kind of movement is needed?

R. Work out

1. Light travels at 300 000 kilometres per second. One light-second is the distance that light travels in one second.
    a. How far does light travel in one second?
    b. What distance is one light-second?
    c. How far does light travel in half a minute (30 seconds)?
    d. How far, in kilometres, is the astronomical distance 30 light-seconds?

2. There are about 30 000 000 (30 million) seconds in a year, so in one year light will travel one million times further than it does in 30 seconds.
    a. How far does light travel in one year?
    b. How far, in kilometres, is one light-year?

Q. Investigate **EXTENSION**

- What explanations do astrologers give for the influence of planets and constellations on our lives?
- How could you test their predictions?

Light emitted by the Sun takes just over eight minutes to reach us. The Sun-Earth distance is eight light-minutes. The distance to the nearest star, Alpha Centauri, is 4.3 light-years. That is, it is about 250 000 times further away than the Sun. No wonder that, from Earth, the Sun looks so different from all the other stars.

Distances in astronomy are so large that special units are used. The average distance of the Earth from the Sun is 149 600 000 km. This is called an astronomical unit (AU). Mars is about 1.5 AU from the Sun, Jupiter 5.2 AU and Neptune 30.1 AU.

Even larger distances are expressed in light-years. Although this sounds like a time, it is actually a distance, as the next activity explains.

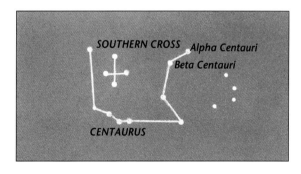

Alpha Centauri and neighbours

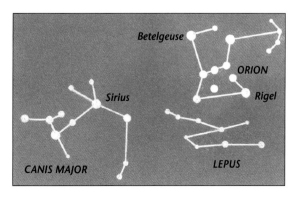

Orion is one of the most brilliant constellations in the sky, with two of the largest stars, Betelgeuse and Rigel. Canis Major is to the east and contains one of the nearest stars, Sirius

Sirius is one of the nearest and brightest stars. It is only 8.8 light-years away. Two other bright stars are much further away than Sirius. Rigel is 880 light-years away and Betelgeuse 520 light-years. They are part of the constellation Orion which is one of the easiest to locate. Look for it in the southern sky on a winter evening.

Stretching across the sky you can see a milky band of thousands of faint stars. This is the Milky Way. To see it you will need a dark, moonless sky. In big towns street lighting makes it impossible to see.

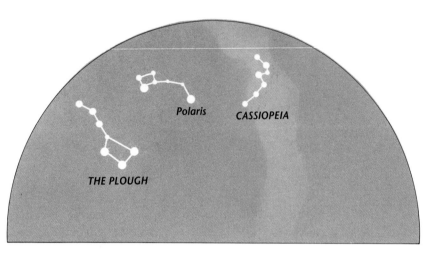

Part of the summer sky showing the Milky Way

When you look at the Milky Way you are looking through a galaxy. The Earth is a very small part of it. (The word galaxy actually comes from the Greek word for 'milk', so galaxy and Milky Way, mean almost the same thing.)

The diagram below shows the galaxy as it would be seen edge-on. The galactic centre lies about 25 000 light-years away from us. When we look along the main plane, many stars are seen in approximately the same direction, and this causes the Milky Way effect.

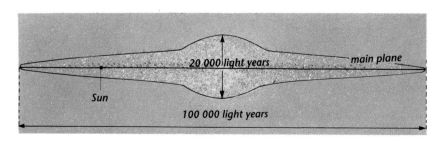

The Sun in our galaxy

The Milky Way is made up of about 100 billion (100 000 000 000) stars. It has a spiral shape, with swirls of stars around a central nucleus. The Sun is towards the outer edge of the galaxy in one of its spiral arms. The Milky Way is about 100 000 light-years across. This spiral galaxy spins about its centre, taking some 220 million years for the Sun to make one revolution. The Sun is actually travelling around the centre of the galaxy at a speed of about 200 km per second.

Spiral galaxies like ours spin relatively rapidly. You can observe the same effect if you pour cream onto the top of a stirred cup of coffee. The great galaxy of Andromeda, which is one of the most distant objects that can be seen with the naked eye, is another rotating, spiral galaxy. It is 2 200 000 light-years from us. Slower spinning galaxies tend to be elliptical in shape. Maffei is an example of this. A third type of galaxy is irregular in shape and not spinning. An example is the Clouds of Magellan.

The Andromeda galaxy

There are millions of galaxies in the universe. They seem to form groups or clusters. There may be over 2000 galaxies in a cluster. The galaxies in our local group, which include the Andromeda galaxy, Maffei and the Clouds of Magellan, are up to 2 200 000 light-years from us.

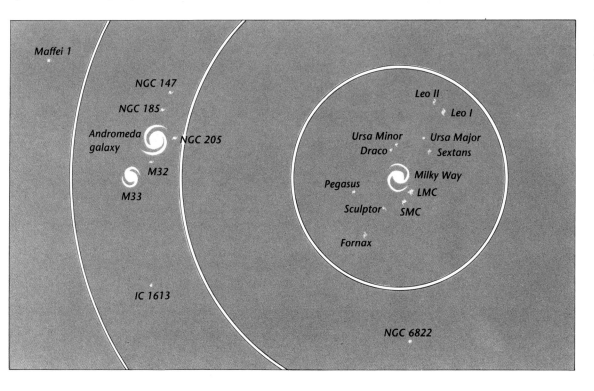

Our Milky Way galaxy is in the local group with its two companions, the LMC and SMC (the large and small Clouds of Magellan)

The most distant galaxy so far detected is 15 000 million light-years away.

S. Work out

- On a piece of paper draw the Sun and Earth. The next star is 250 000 times further away from the Earth than the Sun. Work out where it would be if you placed it to scale.

- Take another piece of paper to represent the universe. (Although we do not actually know what shape the universe is, we can be sure that this is not a very good representation of it. Perhaps you could suggest a better one.) Decide on a symbol to represent a galaxy (you can use different sizes and shapes of symbol if you want to).

  Use the description of the universe in the text, and any additional information, to draw a diagram of the universe. You should include:

  the Milky Way, its size and the position of the Sun;

  the local group of galaxies, including Andromeda, and its distance from the Milky Way;

  other clusters of galaxies;

  the distance to the furthest known galaxy.

  Make notes to explain any uncertainties in your diagram.

  Comment on your problems of showing the universe to scale.

UNIT 5
UNDERSTANDING THE SOLAR SYSTEM

Section 5.1

SCIENTIFIC EXPLANATIONS

The universe according to Ptolemy, from an old print, 1600. The arrangement of the celestial bodies according to the Ptolemaic theory is clearly shown, although no attempt has been made to make the distances even approximately correct

The last unit emphasised how astronomy has been developed by observations. It is difficult to do laboratory experiments in astronomy, as one might to investigate the behaviour of a plant, a chemical reaction or the safety of a car. Scientific knowledge is usually gained by testing ideas (hypotheses) against experimental observations. In this way an explanation (theory) is developed which can be further tested. If scientists cannot easily set up experiments to study the real thing, they may use a model to help develop the theory.

In explaining the origin of the solar system there are obvious problems of experimenting. We would need to travel great distances, and go back in time. A scientist has to act like a detective in solving the problem of a crime. The scientist was not there at the time and place that the solar system developed just as the detective was not present when the crime was committed. Both will observe, collect evidence and analyse information. The detective may stage a reconstruction of the crime to help. The scientist may build a model of the solar system to help explain the way the planets move.

A. Discuss

- It has been suggested that in some areas of science experiments are difficult or impossible to carry out, and models are more important in developing explanations.

  Test this idea by making two lists from topics you have studied in this course or read about elsewhere:

  1 Topics in which experiments are possible, and theories can be tested practically;

  2 Topics in which no experiments can be done, and explanations are tested using models and observations rather than practical experiments.

  Did all the topics you thought of fit easily into one list or the other?

  (Ask for Worksheet EU25 *Testing a theory* if you need help.)

Although scientists cannot do experiments on systems as large as our solar system some experimental work is possible. The Viking mission to Mars (1976) tested rock and soil samples for evidence of life. (No conclusive evidence for any life forms was found.)

The clouds of Venus are now known to be made of sulphuric acid. Laboratory experiments on Earth can help work out how they were formed from chemical reactions between sulphur dioxide and water. This is a kind of practical model. It uses chemical substances but not those from Venus! Many scientific models now use computers.

B. Model

- Use a computer software drawing program to design your own model of the solar system.

  What information did you use? (You could refer to Worksheet EU22 *Planet data*.)

Models and explanations

Mankind has been developing explanations for the origin of the Earth and the universe since the earliest times. A scientific model can be used in the development of a theory. It leads to predictions which can be tested, and thus leads to new knowledge. Models are also used to communicate what we already know and want to explain to others. Many of the models you have met in science lessons are of this kind. They are designed for teaching. Many models of the universe were not designed for scientific use. They are about purpose and meaning; they may include gods or strange creatures. Many are concerned with trying to explain *why* the Earth and universe exist rather than *how*. Here are some examples from the past.

1 Chinese explanations of eclipses (about 2000 BCE)

The eclipse is caused by an unfriendly dragon which is trying to eat the Sun or moon. This can be prevented by making a lot of noise to scare the dragon away. It is important to know when this will happen so you can be prepared to shout and scream at the dragon. Chinese astronomers were able to do this by recognising a pattern in their records of previous eclipses. There is a story that in 2136 BCE the chief astronomers Hsi and Ho failed to predict an eclipse and were executed for neglecting their duties.

2 Jewish account of the creating of the universe (about 600 BCE)

In the beginning when God created the universe the Earth was formless and desolate. The raging ocean that covered everything was engulfed in total darkness and the power of God was moving over the water. Then God commanded 'Let there be light' and light appeared. God was pleased with what he saw.... Then God commanded 'Let lights appear in the sky to separate day from night and to show the time when days, years and religious festivals begin; they will shine in the sky to give light to the Earth' and it was done. So God made the two larger lights, the Sun to rule the day and the moon to rule over the night; he also made the stars.

The Bible, Genesis Chapter 1 verses 1-4, 14-16

This illustration of So God made two larger lights ... *(Genesis 1.16) was painted by the Italian artist Raphael and his assistants in 1519*

3 Egyptian explanation of the structure of the universe (about 1000 BCE)

The universe is a box shape. The sky is held up by pillars at each corner. The Sun and moon are carried in boats which sail along the river in the sky called Ur-nes. Egypt and the river Nile are in the centre of the Earth, which is surrounded by oceans.

Early Chinese model of an eclipse

4 Hindu (Indian) belief in cycles of creation (about 500 BCE)

In the *Bhagarad Gita* the Hindu god Krishna says:

All this visible universe comes from my invisible Being ... I am the source of all being, I support them all but I rest not in them ... At the end of the night of time all things return to my nature; and when the new day of time begins I bring them again into light. Thus through my nature I bring forth all creation and this rolls round in the circles of time. But I am not bound by this vast work of creation, I am and I watch the drama of works.

These cycles of death and rebirth or creation are expressed as the cosmic dance of Shiva. There seem to be some similarities here with very recent ideas about how the universe changes. These are described in the next unit.

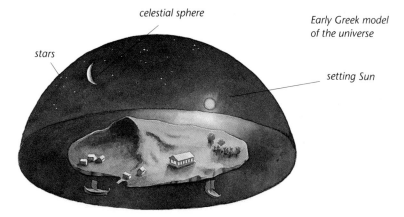

Early Greek model of the universe

Statue of the Hindu god Shiva

C. Think about

- Consider the features of each of the models of the universe.

  1. Which of them might help to give a picture of *how* the universe behaves?
  2. Which models are more about *why* the universe is there?
  3. Which of the models could be tested by experiment?
  4. Suggest a prediction from some of the models and how it could be tested.

5 Greek model of the universe (600 BCE)

The Greeks were very interested in ideas, and made many models of the universe. The earliest was by Thales of Miletus (now in Turkey). He was a successful observer and correctly predicted the date of a solar eclipse in 585 BCE. This was taken as an omen by two warring groups who stopped fighting. His model was that the Earth was flat and floating in water. It was surrounded by a sphere which rotated around the Earth and held up the Sun, moon, planets and stars.

Several of the ideas have a religious importance which cannot be scientifically tested. Others are concerned with accurate predictions which can be tested, but the significance of these predictions is open to question. Some early civilisations turned their detailed observations into useful practical aids without ever, it seems, needing to explain the patterns. The Babylonians of the Middle East (about 700 BCE) and the Mayans of south-central America (about 700 AD) used observations for measuring time and making calenders. The Chinese used their knowledge of the unchanging position of the pole star for navigation.

The Greek civilisation which began around 600 BCE lasted several hundred years and made many important discoveries in astronomy. These were often a combination of ideas or models with careful observation and measurement. They formed the basis of our scientific understanding of the solar system.

D. Research

- Find out, by using encyclopaedias or reference books, about the following scientists and their discoveries, and how much they relied on observations, models and on experiments:

  1 Pythagoras (born 580 BCE on Samos, a Greek island) proposed that the Earth was spherical.

  2 Democritrus (Greek, died 380 BCE) recognised that the Milky Way consisted of many stars and that the moon was similar to the Earth.

  3 Aristarchus (born 310 BCE on Samos) asserted that the Sun was the centre of the solar system and that the planets revolved around it. He estimated that the Sun was eight million kilometres from the Earth.

  4 Eratosthenes (born 276 BCE in Libya) calculated that the circumference of the Earth was 46 000 kilometres.

  5 Hipparchos (born 190 BCE in Turkey) calculated the distance to the moon, the size of the moon and made a star map.

  6 Ptolemy (born in Egypt about 100 AD) combined most of the ideas and knowledge up to this time in a complex model which had the Earth at the centre of many spheres carrying the planets.

Ptolemy's model was so successful in predicting the motions of the Sun, moon and planets that it survived for about 1500 years. In 1514 Nicholas Copernicus of Poland re-established the ideas of Aristarchus and proposed a Sun-centred (heliocentric) view of the solar system. The Earth and planets orbited the Sun in circular paths in this model. Very careful measurements by the Danish astronomer Tycho Brahe were used by the German Johannes Kepler in a series of laws which he published between 1609 and 1618. These described with great accuracy the paths of planets round the Sun. He showed that they are ellipses and not circles.

At about this time the great Italian scientist Galileo had made his own telescope. This was just two years after it was invented by the Dutchman Hans Lippershay. Galileo made important discoveries with his telescope, including the facts that Jupiter had moons and that the Earth's moon had mountains. He was also able to use his mathematics to begin to explain how the planets could move as described by Kepler's laws.

A portrait of Galileo Galilei, the great Italian scientist who was the first great astronomer to use a telescope

There was a lot of opposition to the idea of a heliocentric solar system. Copernicus was afraid he would be ridiculed. Galileo had serious disputes with church authorities including the pope. People had accepted for hundreds of years the model of Ptolemy and Aristotle's explanations of how things moved. What Galileo and others proposed needed a revolution in their thinking.

E. Work out

- How do *you* know that the Earth goes round the Sun and not the Sun round the Earth? Could you persuade someone who had the opposite view?

- The Earth seems to be stationary and the Sun appears to move. What evidence, models and explanations would you use?

- Set up a debate between members of your group holding opposite views. You will need to do some research and organise your arguments.

  Worksheet EU26 *Does the Earth move?* may help you if you are stuck.

Section 5.2

UNIVERSAL GRAVITATION

What makes the planets in the solar system orbit the Sun? Isaac Newton was able to answer this question. He developed the work of Galileo, both in mechanics and astronomy and established the idea of universal gravitation.

You will be familiar with gravity as a force which pulls objects downwards, that is, towards the centre of the Earth. Newton proposed that gravity is universal. All objects in the universe, with mass, attract other objects with mass. Newton was able to explain the orbits of the planets by his theory of gravitation. Gravity is the force which keeps the planets moving in near-circles. The need for such a force (called the centripetal force) is explained in Unit 1 of *In transit.* You can get the idea of how it works by swinging an object round on the end of a string. The string provides the centripetal force to keep the object in orbit.

The force of gravity keeps the moon in orbit round the Earth; and the Earth and other planets in orbit round the Sun. It is also responsible for the much larger scale organisation of the stars into galaxies and the galaxies into clusters. Gravitational forces extend throughout the universe.

A better model to show gravity in the universe is to use a stretched rubber sheet, with objects placed on it to represent stars or planets. This is demonstrated in Unit 2 of *In transit* and is shown in the diagram.

F. Work out

1 How does the model show that:

a the force of gravity is in the direction which joins the centre of the two bodies;

b the force of gravity is greater, the larger is the mass of each of the bodies;

c the force of gravity is smaller, the further apart the bodies are?

(Hint: look at the shape of the surface).

2 Use the following information to explain how the weight of one kilogram mass is different on the moon from on the Earth:

The strength of the gravitational force depends on:

a the mass of the objects which attract;

b their distance apart (if they are in contact, this is the distance apart of their centres).

Newton's theory of gravitation has proved to be very successful in helping us understand the solar system. When William Herschel discovered Uranus in 1781 its orbit matched what you would predict from gravity, almost perfectly. However, in 1821, the French astronomer Alexis Bouvard noticed that there were some differences between the orbit of Uranus and the predictions from Newton's theory.

G. Discuss and research

1 What explanations can you suggest for the differences between the orbit of Uranus and the predictions based on Newton's theory of gravity?

2 Find out how the work of other astronomers on this problem led to the discovery of a new planet, Neptune, by the German Johann Galle in 1846.

H. Work out **EXTENSION**

- From his laws of motion (see *In transit*) and gravitation Newton could explain Kepler's laws of planetary motion. One of these laws states that:

  $\dfrac{r^3}{T^2}$ is constant

  where T is the period of a planet's orbit and r is its average distance from the Sun.

  You can check that this is true from the planet data on Worksheet EU22.

- Make a table for the planets listing T and r for each.

- First work out r^3 and T^2, and then $\dfrac{r^3}{T^2}$ (easier than the other way round!). Check that $\dfrac{r^3}{T^2}$ is constant. (You will need a calculator and to be good with powers of 10!)

- Explain in your own words how you think the orbit of a satellite or a planet depends on gravity.

The moon is called a satellite of the Earth because it is in orbit around it. Nowadays there are many other satellites orbiting the Earth, which have been sent up to explore space, monitor conditions on Earth and provide telecommunications systems. You can learn more about the uses of satellites and how they are given the correct orbits in *Systems and control* Unit 2.

Section 5.3

THE ORIGIN OF THE SOLAR SYSTEM

Since the revolution in astronomy in the seventeenth and eighteenth centuries there have been two main types of theory about the origin of the solar system: nebular theories, and catastrophe theories. According to the first the planets formed from a cloud of gas and dust (a nebula). This may have been at the same time as the Sun formed, or perhaps later from material left over after the Sun had formed.

According to the nebular theories the Sun and planets condensed from a cloud of interstellar gas and dust

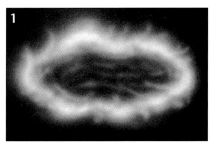

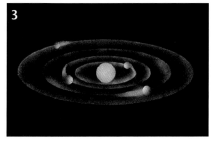
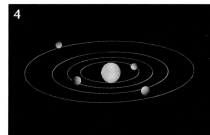

The catastrophe of the second type of theory happened when a second star passed very close to the Sun. The Sun exerted huge forces on it and pulled matter from it. This ejected material became the planets and other bodies in the solar system.

Catastrophe theories account for the existence of planets around the Sun by a close encounter with a passing star

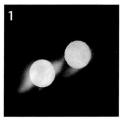

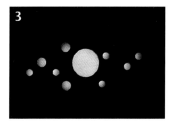

There are several variations on each of these themes, and many have been worked out in considerable detail. At the present time the nebular theory is more widely supported.

The main stages in the formation of the planets according to the nebular theories are:

1. When the Sun was new it was much hotter than it is now so it evaporated the parts of the nebula near to it.

2. Gravity pulled the solid particles in the nebula together to form planets which heated up inside.

3. The evaporated nebula formed a hot gas which blew outwards from the Sun as a solar wind. This affected the inner planets more than the outer ones. In time this solar wind died down and it is now very weak.

4. The surfaces of the planets cooled as the Sun cooled, but the insides remained hot.

5. Planets can form atmospheres from chemical reactions at their surface, and when volcanoes bring up new materials.

The composition of planets is now quite well known. Brief details of the atmosphere, surface and core were given in Worksheet EU22. If you look at this you will see that their compositions are given as rocky or icy. Rocky substances have high melting points; they include metals, oxides and silicates. Icy substances have low melting points; they include water and gases such as oxygen, nitrogen, methane, carbon dioxide and ammonia. In many planets the temperature is low enough for these to be solid. Hydrogen and helium have the lowest melting and boiling points, so they are often gases.

Life on Mars?

People have often thought that there might be life on Mars: books and films have been made about aliens from there. One hundred years ago a famous astronomer Percy Lowell was convinced that Mars was covered with canals. Later another scientist pointed out that Mars was dry and colder than ice! Nevertheless there was great interest in 1976 when the Americans sent the Viking expedition to try to find out if life existed there.

Viking lander module taking samples of the surface of Mars, August 1976

The landing craft analysed samples of soil and sent back many excellent pictures and other data. Much is still being analysed. The overall conclusion about life – probably not.

I. Work out

- Using the information in the previous paragraph and the explanation of the solar system's origin:

  1. Separate the planets into those which are rocky and icy/rocky.
  2. What general rule can you write about these?
  3. The Earth and Mars have icy substances on their surface nowadays but not in their insides. Why do you think this is?
  4. Why does Mercury have no atmosphere?
  5. Where could the atmosphere of Venus have come from?
  6. There was once lots of hydrogen and helium throughout the solar system. Where is it now? Why do think this is?

J. Plan and present

- Plan some experiments that you would carry out on the next expedition to Mars, to try to be certain that no life exists there.
- Review the information you already have about Mars and about the conditions needed for life. You will probably need to do some more research.
- Present your plans as a poster or a talk to your class.

Section 5.4

HOW LONG WILL THE SUN LAST?

For a brief period under the pharaoh Akhenaten (about 1350 BCE) the ancient Egyptians worshipped the Sun as the only god. The god, Aten, is represented as the solar disc, with rays ending in hands

Our very existence here on Earth depends on the Sun continuing to shine and deliver energy to our planet. It is not surprising that scientists are working hard to understand the Sun and predict its future behaviour. For how long will it go on shining? When change comes, what will it be? This section and the next unit are about these questions.

The effects of a change in the Sun's output on conditions here on Earth is a theme used by several science fiction writers

To understand how the Sun might change in the future it is essential to know its present rate of energy output.

K. Discuss

1 Suppose you have an instrument that measures the rate at which it receives solar energy. You want to use this measurement to calculate the total rate at which the Sun is radiating energy. Under what conditions would you take measurements with your instrument?

2 Your instrument gives a value for the rate at which energy is arriving at the Earth's surface, in mW/cm² (or W/m²). What other information would you need in order to calculate the total power radiated by the Sun?

3 What unknown and uncontrollable factors might affect your estimate of the Sun's power?

A simple solar cell made of silicon responds to the light falling on it. A direct measurement of current from the cell can be used to estimate the total light intensity falling on it.

L. Observe

- Record the current produced by a solar cell when it is exposed to different kinds of illumination. Check that the current depends on how bright the light is.

  How does the current vary as the cell is moved away from and towards a lamp?

  What current is produced when the cell is in direct sunshine?

  What is the current when it is in the shade?

  What difficulties and uncertainties are there in using a measurement of the solar cell's current to estimate the intensity of solar radiation, and hence the power of the Sun?

A typical solar cell is more sensitive to some colours than to others.

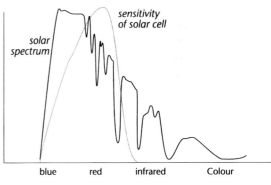

Nearly half of the radiation from the Sun is absorbed, scattered or reflected by the gas molecules and other particles in the atmosphere. An average value for the intensity above the Earth's atmosphere is 1.4 kW/m² (enough to heat water for a cup of tea in less than a minute, if you could collect and use all the energy arriving in the square metre). Energy is crossing every square metre at this distance from the Sun whether the Earth happens to be there or not. To find the total rate at which the Sun is radiating energy we need to know the total area at this distance. This is the area of a sphere at the Earth's distance from the Sun.

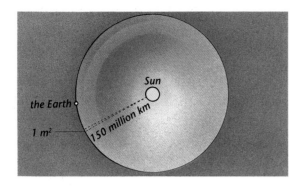

The Sun's energy source is not a fossil fuel like coal or oil, but nuclear fuel. Our present nuclear power stations depend on nuclear fission. Fission means splitting. Nuclear fission is the process in which the nucleus of a heavy atom, like uranium, splits into smaller fragments with the release of energy. The fission of one kilogram of uranium-235 (a rather rare isotope of uranium) releases 7×10^{13} J (70 million MJ).

The Sun's energy source is not nuclear fission but nuclear fusion. Fusion means joining together and nuclear fusion is the joining of two light nuclei to form a heavier one. The fusion of one kilogram of deuterium (heavy hydrogen) to form helium releases 6×10^{14} J (600 million MJ).

Nuclear fusion

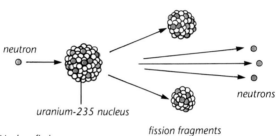

two hydrogen nuclei → helium nucleus

The Earth is 150 million kilometres (150×10^6 km, or 150×10^9 m) from the Sun, so the surface area of the sphere shown in the diagram is about 3×10^{23} m² (3 followed by 23 zeros).

Thus our estimate of the total rate at which the Sun is radiating energy is
1.4 kW/m² × 3×10^{23} m²
= 4.2×10^{23} kW or 4.2×10^{26} joules per second.

This is such a colossal rate of energy output that we are bound to ask 'How long can it go on at this rate?' Yet we believe that the Sun is about 5000 million years old, and that there has been life of some form on the Earth for about 4000 million years; plant life for perhaps 1300 million years and animal life for 700 million years. Presumably the Sun has been radiating at something like the present rate for much, if not all of this period.

Where does the Sun's energy come from?

Nuclear fission

neutron → uranium-235 nucleus → fission fragments, neutrons

Hydrogen fusion is the basis of the hydrogen bomb, and taming hydrogen fusion for the controlled release of energy to power electrical generators has kept many scientists and engineers busy for much of the second part of the twentieth century. 60% of the Sun's mass is hydrogen; the next most abundant element is helium, the product of hydrogen fusion. One of the factors that makes hydrogen fusion so difficult to control on Earth is that it occurs at extremely high temperatures. This makes it difficult to confine the fusing hydrogen in any kind of container. The temperature inside the Sun is estimated to be about 10 000 000°C.

> **M. Work out**   **EXTENSION**
>
> - The Sun is huge. Perhaps it is so big that it is able to continue to pour out energy by a fairly simple chemical process such as burning coal or oil, and yet not use up enough for us to notice if getting smaller. Check this using Worksheet EU29 *What is the Sun's fuel?*

Fairly straightforward calculations (but involving very large numbers), like those in the first part of Worksheet EU29, rule out the possibility that the Sun is powered by a fuel like oil or coal. George Gamow once calculated that if the Sun was like a coal fire (with a plentiful supply of oxygen) which was lit 5000 years ago, the fire would be burnt out by now!

Calculations like those on Worksheet EU29 *What is the Sun's fuel?* show that nuclear fusion of hydrogen to helium can keep the Sun going, more or less as at present, for 10^{11} years (100 000 million years). This is comfortably longer than the estimated age of the solar system. As well as forming helium from hydrogen, further fusion reactions give rise to the heavier elements like carbon, oxygen and nitrogen which life depends on, and all the other elements, including the metals.

UNIT 6
BACK TO THE BEGINNING

Section 6.1

STAR TYPES

The Sun is powered by nuclear fusion reactions that will allow it to radiate energy at the present rate for 100 000 million years or so. To us the Sun looks different from all the other stars. In fact the Sun is a very typical star; it looks different because we are so very much closer to it.

Because the distances they have to work with are so huge, astronomers often use special units such as the 'light-year' (see Section 4).

> **A. Think about**
>
> - Alpha Centauri and Sirius are the Sun's nearest neighbours. Both these stars look bright, but much more distant stars can also appear bright. What, apart from distance, might affect how bright a star appears to us?

Alpha Centauri is the bright star on the left

The constellation Canis Major (Big Dog). The main star here is Sirius, the brightest star in the sky

Star	Brightness (relative to Sun)	Distance (light-years)
Sun	1	0.000 017
Sirius	23	8.8
Alpha Centauri	1.7	4.3
Rigel	50 000	880
Betelgeuse	14 000	520
Polaris (pole star)	8 000	820

In this table the stars are listed in order of their apparent brightness, as seen from Earth.

Rigel is one of the brightest stars we can see – not because it is nearby, but because it actually is a very bright star. On a scale which gives the Sun a brightness of 1, Betelgeuse and Rigel are much brighter than the much closer stars Alpha Centauri and Sirius. We would probably not notice a star like the Sun if it was the same distance away as Rigel or Betelgeuse.

To be able to understand stars better we need to be able to say how bright a star 'really' is, that is, how bright it would seem if it was as close to us as the Sun is. To do this we have to know how far away it is.

How far are the stars?

Estimating the distance of all but a few of the nearest stars posed astronomers with a real challenge. Vital work was done by Henrietta Leavitt at the beginning of the twentieth century. She worked at Harvard Observatory, USA, where she had the tedious-sounding job of cataloguing star photographs. Among the stars recorded on these photographic plates were some 'variable stars'; stars whose brightness changes in very regular cycles. Different variable stars have different cycles, ranging between about two and 100 days.

Some of these variable stars are actually in another galaxy. This means that they are so far away that the small differences in distance from us do not affect how bright they seem. Leavitt noticed that the brightest of these variable stars take longer to go through their cycle. Now our observation of a variable star's cycle is not affected by how far away it is, so we can be sure that there is a relationship between a variable star's cycle and its real brightness. Leavitt's discovery was taken up by other astronomers, who were able to get independent estimates of the distance to some relatively nearby variables in our own galaxy. This was only one step in the complex business of measuring the universe, but much of the astronomy that has been developed since then depends on Leavitt's work.

B. Work out

1 What is meant by the term 'variable star'? Sketch a graph which shows how the brightness of two different variable stars might change with time.

2 Why is it important that Leavitt's observations were made on very distant variable stars? Explain why the brightness of a star might depend on its distance from us, but the cycle of its variation does not.

3 What subsequent steps are necessary in using Leavitt's discovery to establish a distance scale for the universe?

4 (Extra) Sketch a graph showing the relationship which Leavitt observed between brightness and period for variable stars.

Red giants and white dwarfs

Even without using a telescope we can see that stars look different. Some are brighter than others, and this may or may not be because they are nearer to us. If you compare Sirius, Rigel and Betelgeuse you will notice other differences. Betelgeuse looks distinctly reddish compared to the other two. Naked-eye observations are not very reliable. By studying the spectra of the light emitted by different stars astronomers have been able to group stars into several different types: Betelgeuse is a 'red supergiant'; Rigel a 'blue supergiant', and Sirius a 'normal' yellow star, rather like the Sun. 'White dwarf' is another star type.

Henrietta S Leavitt (1868-1921)

Much information is gained by studying the spectra of light emitted by stars

The colour of light emitted by any hot object depends on its temperature. We recognise this when we talk about something being red hot or white hot. The higher the temperature the greater the proportion of blue light (short wavelength) that it emits; so blue stars are the hottest and red stars the coolest. The surface of a blue star may be 25 000°C, and a red one only 3000°C. The Sun's surface is about 6000°C, quite typical for a normal star. Most stars are neither giants nor dwarfs, but even so-called normal stars can vary in colour from red to blue, with surface temperatures from 3000°C to 35 000°C.

The names 'red giant' and 'white dwarf' tell us something about the size of a star, as well as its temperature. Radius and mass are among the properties of stars which astronomers are now able to estimate; not by direct observation and measurement, but by careful interpretation of indirect evidence. The size of nearer stars can be measured. Their composition can be worked out from studying the spectra of their light. From the size and density of a star the mass can be worked out.

C. Research   EXTENSION

- Use books on astronomy to find out how astronomers are able to calculate the masses of distant stars.

  (Hint: look for binary stars)

The constellation of Orion. The bright, pinkish star at upper left is Betelgeuse; the bright star bottom right is Rigel

Section 6.2

BIRTH AND DEATH OF A STAR

By combining information about the variety of types of star with their different temperatures, sizes and brightnesses, together with ideas about gravity and nuclear fusion, scientists have been able to work out the life history of a typical star, like the Sun.

The story starts in a swirling cloud of gas (mostly hydrogen) and tiny particles. The particles are not spread out smoothly; parts of the cloud are more dense than others. A star begins to form as gravity pulls the cloud together at its denser points. As the particles are pulled in, collisions between them heat up the 'protostar'. This gravitational collapse raises the temperature enough to cause the star to start to glow. The temperature continues to rise until, at the core, it reaches millions of degrees. Hydrogen fusion starts. As the temperature rises further the rate of hydrogen fusion increases, and the star achieves a kind of balance. The inward pull of gravity is balanced by the pressure of the very energetic particles in the star's hot core. This is the state the Sun is in now, and has been for perhaps 5 000 000 000 years. It will continue like this for another 5 000 000 000 years or so. More massive stars contract more rapidly, rise to higher temperatures and glow more brightly as hydrogen is converted to helium more rapidly.

These are the blue stars, whose stable period may last only 500 000 years. The rule here is 'the bigger, the brighter – and the briefer'.

Eventually, even for a comparatively cool and slow star like the Sun, the hydrogen in the core gets used up and the rate at which energy is released there slows down. Collapse starts again, leading to even higher temperatures at the core, and release of energy from hydrogen fusion in a shell surrounding the core. This causes the star to swell. As it does so the outer layers, further from the nuclear furnace, cool. The star has become a red giant. At this stage (not due for billions of years!) the Sun will engulf the Earth. A more massive star becomes a red supergiant, like Betelgeuse.

Massive star forms nebula out of cloud of dust and gas

After a few million years star begins to shine

Because star is so massive the red-giant stage is reached much more quickly than by a star like the Sun

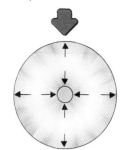
Star is too massive to survive and collapses under its own gravity

Star's outer layers are blown into space within seconds of the beginning of the collapse

All that remains is a small, super-dense star and an expanding cloud of gas

The core collapses, bringing about higher temperatures, up to 100 000 000°C. Now even stable helium nuclei start to react. Three can be squeezed together to form carbon:

$^4He + {}^4He + {}^4He \longrightarrow {}^{12}C$

Oxygen and neon are also formed. Other reactions producing the nuclei of heavier elements may occur, particularly in more massive stars. This stage of a star's life is relatively brief, and it fairly soon becomes a white dwarf. As the remaining nuclear fuel is used up, a star which spent most of its life like the Sun, slowly cools and contracts until it is the size of the Earth, although much much more massive.

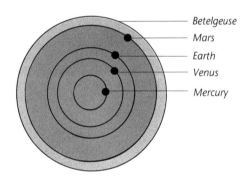

More massive stars have a different end. Such a star has to shed excess matter before its final collapse. Very occasionally a huge 'explosion', called a supernova is observed in the sky. This is the final outpouring of energy and matter from a large star. Such events are rarely seen. The most recent one in our galaxy was in 1604. A very spectacular one happened in 1054. Its remains can be seen today.

The 'crab nebula' is the remains of the supernova of 1054

The 1054 supernova was seen and recorded in China where it caused much consternation

> ### D. Work out
>
> - Use the information given in this section about the birth and death of a star to answer the following questions:
>
> 1 Is the Sun:
>   a about half-way through its life cycle;
>   b nearer the beginning of its life;
>   c nearer the end of its life?
>
> 2 a Is the Sun a red giant, blue giant, white dwarf or a 'normal' star?
>   b Which other stages will the Sun go through in the future?
>
> 3 A red star is cooler than a yellow star, and yet a red supergiant (like Betelgeuse) can be brighter than a yellow star. Explain.
>
> 4 Why are there relatively few blue giant stars? (Think about the length of time that a star stays in this state, compared with 'normal stars' like the Sun.)

At the very end of a star's life, the star may have become so compact that gravitational forces are strong enough to force all the nuclei together to form an incredibly dense 'neutron star'. The mass of two Suns would form a ball about 50 kilometres across, with a density 100 million million (10^{14}) times that of the Sun. Pulsars, objects that emit very regular bursts of radio waves, are thought to be rotating neutron stars.

An even more strange fate may await just slightly more massive stars. If it is more than three solar masses the star continues to collapse and the gravitational force becomes so strong that nothing, not even light, can escape from it. It may sound like science fiction, but scientists are convinced that such 'black holes' can exist. So far their existence has not been proved, but astronomers are beginning to get indirect evidence for them.

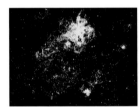
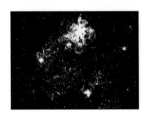

Supernova 1987a before and after eruption

Section 6.3

THE EXPANDING UNIVERSE

Edwin Hubble (1889 – 1953)

In 1919 Edwin Hubble, a 30 year-old astronomer, joined the Mount Wilson Observatory in California where a new telescope with a 2.5 metre mirror had just been installed. This '100-inch' telescope remained the world's largest until a mirror with double the diameter, the '200-inch', came into operation in 1948.

Ten years later Edwin Hubble announced a discovery that was to revolutionise astronomy; the galaxies are moving away from us, and the further away they are, the faster they are moving. His conclusion was based on indirect evidence; there is no way of measuring directly either the speed of a galaxy or its distance. He worked out the speed using the effect we all notice when a police car or ambulance goes past with its siren on.

The giant 4.2 metre diameter mirror of the William Herschel telescope undergoing its final polishing

F. Observe

- Listen to the different sound you hear as a source of sound is coming towards you and as it moves away from you. When is the pitch higher and when is it lower?

- Higher pitch means higher frequency and shorter wavelength. Is the wavelength you detect from a source moving away from you longer or shorter than for the same source at rest?

- Red light has a longer wavelength than blue light. Would you expect a light source that was moving away from you to look redder or bluer than normal?

E. Work out

1 The amount of light collected by a telescope depends on the area of its mirror. Compare the amount of light collected by a mirror with a diameter of 200 inches to that collected by a 100 inch diameter mirror. Is it:

   a the same;

   b double;

   c four times larger;

   d eight times larger?

2 The diameter of the mirror in the telescope with which Herschel discovered Uranus was six inches. How much more light than this telescope does the 200 inch mirror collect?

3 What difficulties might arise in making and in using such a large telescope?

The fact that wavelength seems to be affected by any movement of source or receiver is well known. It is called the Doppler effect, in honour of the Austrian who investigated it for sound in 1842. The effect is much more obvious for sound than for light because sound travels much more slowly than light; their speeds are 330 m/s and 300 000 000 m/s respectively. A police car travelling at 30 m/s (about 110 kph, or 67.5 mph) is travelling at about one-tenth the speed of sound, and we notice the change in pitch of its siren as it goes past. To get a comparable effect with light the source would need to be moving at 30 000 000 m/s (108 000 000 kph or 67 000 000 mph).

G. Discuss — EXTENSION

- A scientist was in court, charged with not stopping at a red light. His defence was that because he was going towards the light wavelengths became shorter and red appeared green. What did the judge, who also knew some science, have to say?

H. Discuss

- Hubble worked out the distances of galaxies of the same type by comparing their brightness. How would he do this and what assumptions would he have to make?

- Hubble worked out the speeds of galaxies by measuring the change in the spectrum of light – the red shift. How would he do this and what assumptions would he have to make?

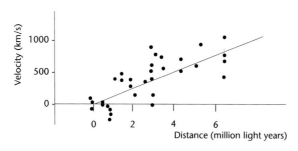

Edwin Hubble's early results, 1929

The reproduction of Hubble's early results show that a certain amount of faith can be useful in interpreting scientific observations! Within two years the observations had been extended to much greater distances and seemed much more secure.

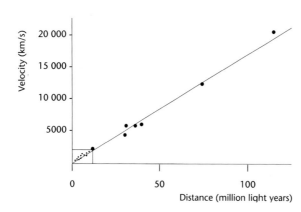

Within two years many more results confirmed Hubble's findings

The findings are summarised in 'Hubble's law', which states that the speed at which a galaxy is moving away from us depends on its distance from us. Hubble's law can be put very simply in symbols:

$v = Hr$

If v is the galaxy's speed in km/s, and r its distance in millions of light-years, then Hubble's constant, H, has the value of 15 km/s for each million light-years. In other words, a galaxy is about a million light-years away if it is moving from us at 15 km/s. The higher the speed, the greater the distance.

I. Work out — EXTENSION

1. The red-shift of light from a galaxy in Ursa Major (the Great Bear or Dipper) shows that it is receding at 15 000 km/s. How far away is it?

2. How far away are galaxies which are receding at 400 000 km/s and 600 000 km/s respectively?

3. The most distant objects that astronomers have been able to detect are thought to be 18 000 million light-years away. How fast are they receding?

The same in all directions

It makes no difference in which direction astronomers point their telescopes: with very few exceptions, the light from all galaxies is red-shifted. This implies that all are moving away from us. The simplest explanation would be that everything else in the universe is moving away, while we remain fixed at the centre. On this picture the Earth (or better, our galaxy) would be like a fixed point inside a balloon being blown up around us. In earlier times mankind might have interpreted the results like this. In the modern view mankind does not have any special central role or position in the universe. We imagine the whole universe, including our galaxy, to be expanding in all directions. Whichever way we look, distances to other galaxies are increasing.

Imagine yourself on a sheet of rubber which is being stretched in all directions. You do not have to be in the middle of the sheet to see everything else moving away from you.

J. Investigate

- Watch as a two-dimensional model of the universe expands.

    What happens to the distance between a pair of galaxies? What happens to the direction from one galaxy to another?

    Are these changes the same for all pairs of galaxies?

    Would an astronomer's observations, and her interpretation of them, depend on which galaxy she was in?

- You can investigate an even simpler model of the expanding universe more carefully for yourself. Worksheet EU31 *The expanding universe* gives details.

    The model is not exactly like the thing it represents. What are some of the differences between these models and the real universe?

- Can you suggest any other, better models for the universe?

To visualise the expansion of the universe you have to imagine space expanding in three dimensions, instead of the rubber being stretched in two.

Section 6.4

THE BIG BANG

Another key piece of evidence about the nature of the universe was discovered more or less by accident in the 1960s. Scientists in the USA were using some very sensitive telecommunications equipment to detect radio emissions from different parts of the sky. They found that the system was plagued by noise – like the hiss you can hear on a badly-tuned radio. All their attempts to get rid of the noise failed. It is said that they even went to the lengths of cleaning out pigeon droppings from the horn that was picking up the radio waves from distant galaxies. To no avail. The birth of the universe itself, rather than pigeons, was responsible.

Theory had predicted, at least 20 years earlier, that some such 'noise' should exist everywhere in the universe. According to the Big Bang theory, the universe began in a huge fireball, about 15 000 million years ago. The temperature may have been as high as 100 000 000 000 (100 billion) degrees. Ever since then the universe has been expanding, and its composition changing. Radiation from the immensely hot fireball has been expanding with the universe ever since. Because of this expansion the radiation is now very much weaker and the wavelengths have been 'stretched'. Instead of the short wavelengths, these radiations are now microwaves, with wavelengths similar to those used in telecommunications (and cookers!) They are what you would expect from something with a temperature of only 3 K (that is, three degrees above the absolute zero of temperature).

When experimenters and theorists met, it soon became clear that the radio noise was not an annoying nuisance – it was an important discovery. More measurements confirmed that the 'noise' matched the radiation expected for a relic of the Big Bang.

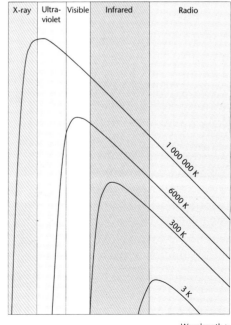

The range of wavelengths emitted by a hot body depends on temperature

A Nasa spacecraft has detected echoes of the galaxies' birth fourteen thousand million years ago. The discovery about the formation of the stars after the Big Bang has been hailed by excited scientists as the Holy Grail of cosmology. **Susan Watts** and **Tom Wilkie** report

How the universe began

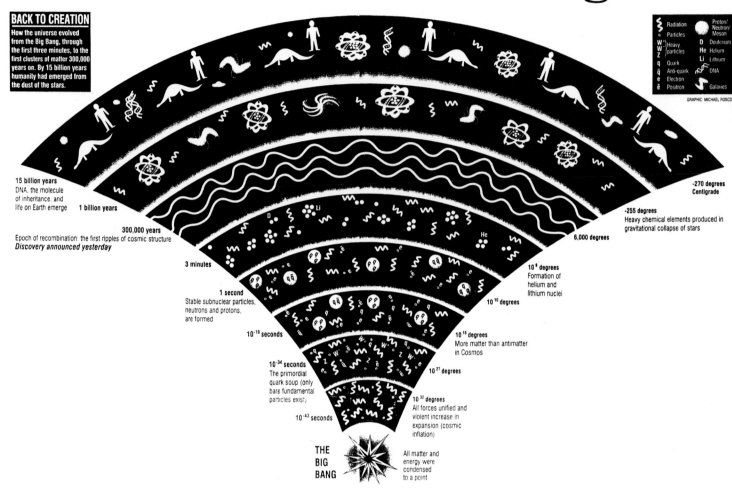

The fact that this 'background radiation' seemed to be completely uniform was rather disturbing. Uniform radiation meant that the universe itself was completely uniform. If the universe had no irregularities how could the present structures like galaxies and clusters have developed? This riddle may have been solved in 1992 with the detection of slight 'ripples' in the background radiation. These are seen as evidence that structures were beginning to form some 300 000 years after the Big Bang. Even before this discovery could be confirmed, it was hailed as a major advance in our understanding of the early universe.

K. Interpret

The newspaper article gives one interpretation of the first 15 billion years of the universe. According to this view:

1. How soon after the Big Bang were the first atomic nuclei formed? What was the temperature at this time? What particles existed before that?

2. When were the first galaxies formed?

3. The diagram seems to show that DNA, dinosaurs and humans all emerged at the same epoch. Is this reasonable?

4. The temperature 15 billion years after the Big Bang is shown as -270°C, but neither humans nor dinosaurs could survive such low temperatures. How should these temperatures be interpreted?

5. One component has been present from the earliest times? What is it?

Section 6.5

THEORIES OF THE FUTURE

It is now generally agreed that the universe began in a cosmic fireball, the Big Bang, about 15 000 000 000 (15 billion) years ago. Scientists have developed a detailed picture of what happened in the very first instants. A book published in 1977 called *The First Three Minutes* describes events up to a time three minutes and 46 seconds after the Big Bang. It gives the so-called 'standard model', an account that starts one-hundredth of a second after the beginning of time, $t = 0$. Since that book was written theorists have pushed back closer and closer to the moment of creation, $t = 0$. A more recent idea is known as 'inflation', and it gets to within 10^{-43} seconds of $t = 0$. (That number is 0.000 000 ... 01 with altogether 42 zeros between the decimal point and the one!) One of the aims of such theorising is to explain how the universe could have developed into the form we know now. It may also have something to say about the future.

One of the biggest unresolved problems of present day cosmology is whether the universe will continue to expand for ever.

> **L. Discuss**
>
> 1 The universe may go on expanding for ever, or it may, sooner or later, start to contract. What might cause the expansion to continue for ever?
>
> 2 What might cause the universe to start to contract, eventually?
>
> 3 What will decide which of these two will actually happen?
>
> 4 It is not possible for the universe to continue for ever at its present size (or any other). Why is this?

The gravitational force between galaxies tends to pull them together, slowing down the expansion caused by the initial Big Bang. The ultimate fate of the universe will depend on the balance between continuing expansion and gravity. We do not yet know just what the outcome will be.

> **M. Observe**
>
>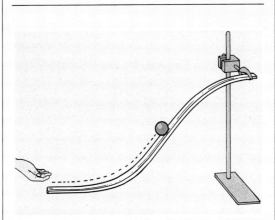
>
> - Watch as a ball is rolled up the 'hill'.
>
>   What factors decide whether the ball will be able to reach the top of the hill?
>
>   What decides whether the ball will roll back down the hill?
>
> - Explain how this model helps us to think about the expansion of the universe?

There are three possible outcomes of the effect of gravity on the expansion of the universe, depending on how much matter there is in the universe. If there is enough matter, gravity will eventually halt and even reverse the expansion, and the universe will start to collapse. However, a universe with less matter in it would go on expanding for ever, although at a slower and slower rate as time passes. These two possibilities are called 'closed' and 'open' universes respectively.

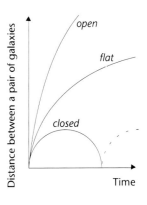

Open, closed and flat universes

Of course there are many possibilities for each, depending on the relative effect of the initial explosion and the opposing force of gravity. The more matter there is in a closed universe the sooner it will begin to shrink. Between the two there is just one third possibility, the so-called 'flat' or 'critical' universe.

> **N. Think about** — EXTENSION
>
> 1 What determines whether the universe is open or closed?
>
> 2 What information do we need to help us decide between the two possibilities?
>
> 3 What observations could give us this information?

If the density of matter in the universe now exceeds a certain critical value, then the universe will, eventually, start to contract. Now this critical density is extremely low, corresponding to an average of only about 10 hydrogen atoms in every cubic metre of space. The amount of matter that can be seen in the stars, galaxies and luminous nebulae, would, if spread out uniformly, be even less than this: the average density of the universe seems to be about 1/100th of the critical value. The hunt for 'cold, dark matter' is on. It is difficult, but not impossible, to detect; and astronomers, using X-ray and infrared detectors for example, are finding more and more evidence for its existence. Black holes could make a significant contribution too.

Any information about how the rate of expansion has changed over time would greatly help to decide between the open and closed possibilities. There is a way in which observations made *now* give information about events and conditions long ago. Remember that light from the Sun takes eight minutes to reach us. Light from the *nearest* star left that star four years before it gets to Earth. Now the red-shift of light from a distant cluster of galaxies gives information about the relative speed *when the light was emitted by the cluster*. Red-shift measurements have been made on clusters more than 1000 million light-years away. So we can infer how fast they were moving 1000 million years ago. The points on the figure show how the speed varies with distance. The three lines that have been drawn on the graph show how we should expect the speed to vary with time on the three different models: a) open; b) flat; c) closed.

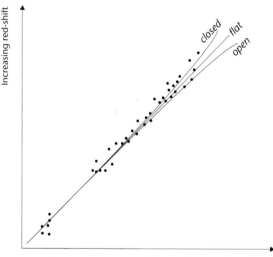

Observations cannot (yet) tell us whether the universe is open or closed

It is pretty clear from the graph that more evidence is needed, especially from even more distant sources, before we can say whether the universe is destined to go on expanding for ever, or will eventually begin to contract.

If the universe is closed, what will happen as it collapses and becomes ever more dense and compact? Will there be a 'Big Crunch'? Will this lead to another Big Bang? If so will the universe go on following this pattern indefinitely? This too may sound like science fiction, but such are the speculations of modern scientists.

The end of science?

The universe seems to be delicately balanced between being open and closed. Perhaps this is just a coincidence. But perhaps not. Suppose the universe were very different. For example, in a universe that went from Big Bang to Big Crunch quickly there would not be time for life as we know it to develop. This would be the case if the force of gravity were stronger – not because there was more matter around, but if the standard force between two one-kilogram lumps of matter one metre apart happened to be larger than it actually is. It turns out that the development of the universe is very sensitive to the value of several physical constants like this. The universe is in a sense very 'finely tuned' to allow, or perhaps even to ensure, the development of life and the evolution of human beings. This has led some scientists and philosophers to suggest that we are, after all, very much part of the 'grand design', and that the universe is only here because we are here to observe it. This goes under the name 'Anthropic principle', and is certainly not accepted by all scientists.

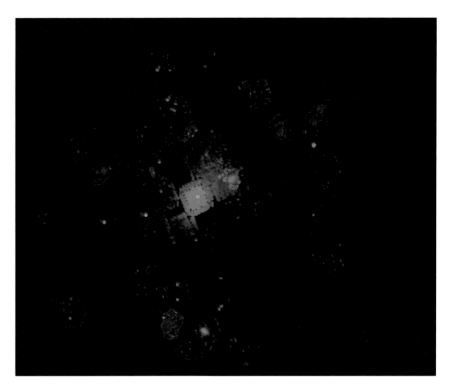

Radio (red) and X-ray (blue) emission from a cluster of galaxies

There have been huge advances in our understanding of the universe in recent years.

New techniques such as radio and X-ray astronomy, and instruments aboard Earth-orbiting satellites have allowed us to see much further out into space, and so back into time. Theoretical physicists have used knowledge about nuclear particles to explain the processes inside stars and in the very earliest stages of the Big Bang. The study of the largest and smallest systems, known or imagined, come together here. Ever more powerful computers are increasingly used to work out the consequences of new theories and models.

At least one scientist has said that he thinks we now have, or are close to having, a 'theory of everything': that science is now, in principle, very close to being able to account for every-thing we know about the universe we live in. Other scientists might argue that our search for knowledge is like a set of Russian dolls. Although every question that is answered increases our understanding, it also raises a new set of questions.

Some people might accept that science may, eventually, be able to answer all our questions about the material world, but this would still leave large areas of knowledge and understanding untouched, simply because they are not the realm of science.

William Blake's illustration of John Milton's lines: 'In his hand he took the golden compasses, prepar'd in Gods Eternal store, to circumscribe This Universe, and all created things...'

Albert Einstein said:
'The most incomprehensible thing about the universe is that it is comprehensible.'

Douglas Adams said:
'There is a theory which states that if ever anyone discovers exactly what the universe is for and why it is here, it will instantly disappear and be replaced by something even more bizarre and inexplicable.'

Issac Asimov said:
'The true delight is in the finding out, rather than in the knowing.'

Stephen Hawking said:
'If we do discover a complete theory of why it is that we and the universe exist, it would be the ultimate triumph of human reason – for then we would know the mind of God.'

O. Discuss

- What do the quotations in the boxes on the left mean? Do you agree with them?
- Use examples from the science you have learnt to discuss the following questions:

  1. Do you think there can be an 'end to science'?

  2. Will we ever get to the point where there is no more to be discovered?

  3. Will there one day be a theory (or theories) to explain everything, or do you think that new discoveries and theories also raise more questions?

  4. Do you think there are some important questions and issues which science will never be able to answer?

Albert Einstein (1879 – 1955)

Index

absolute dating 69
acid rain 60
air, fractional distillation of 12
air masses 24
andesite 73, 91
anticyclone 27, 28
anthropic principle 124
asteroids 100
asthenosphere 91
astrologers 103
astronomical unit (AU) 103
astronomy 95
atmosphere
– changes and weather 23–24, 26–27
– composition 4–5
– evolution of 8
– layers of 6–7
Bacon, Francis 85
baked zone 78
basalt 71, 83, 92
batholith 71
bed, bedding plane 68
Big Bang 121–123
Big Crunch 124
black holes 118, 124
Bode's Law 99
boulder clay (till) 63
carnivore 39
chlorofluorocarbons 10
chlorophyll
clastic sedimentary rocks 66
climatic change 95
clusters (of galaxies) 105
coal 67–68, 94–95
combustion 12
comet 101
constellations 96, 102
continental drift 85, 86
continental shelf 85, 87
convection currents 93–94
core, Earth's 64, 82
cross-cutting relationships 68
crust 82–93
crystals, in rocks 72
decay 40
decomposer 40–42
deformation, of rocks 75–77
density 82–83
deposition 58, 61
depression 26–28
deserts, spreading of 53
diffusion 35
Doppler effect 119
Earth: layered structure 82
earthquake 80–81
– epicentre 80
– focus 80
Earth's magnetic field 82
Earth's resources 95
eclipse 107
ecosystem 32–35
endothermic 20
energy, global balance 22–23
– in ecosystems 38–41
erosion 58, 61

evaporites 67, 95
exosphere 7
exothermic 20
expanding Universe 120–124
extinction 41, 94
faults 68, 76–77
fold 68, 76
– anticline 76
– syncline 76
fold mountains 77, 85, 91
fossiliferous 66
fossils 68, 76
freeze-thaw 59
fronts 27
frost shattering 59
fusion 114, 117
gabbro 70
galaxy 104, 120
– clusters of groups 105
Galileo 109
game reserve 37
geological resources, in Britain 95
geological time 68, 70
glaciers 63
glaciation 85
Glossopteris 86
gneiss 79, 93
granite 60, 71, 72, 75, 83, 93, 95
Greek astronomy 108–109
greenhouse effect 23–24
group (of galaxies) 105
gravitational collapse
– of stars 117
– of Universe 123
gravity 110
half-life 69
hedgerow: removal 47–48
heliocentric 109
herbivore 38
Herschell, William & Caroline 97–98, 100
Hubble, Edwin 119
Hubble's Law 120
hydration 59
hydrogen bonding 21
hydrological cycle 14
ice, rock formation bu 63
igneous rocks 60, 68, 70, 71–74
ionsphere 7
island arcs 88
isotope 69
kaolinite 58
land: use and management 44–52
landscape
– British 45
– farming 46
– formation 70
lava 71
Leavitt, Henrietta 116
light year 103
limestone 60, 67, 95
lithification 66
lithosphere 91–92
magama 71, 92
– chamber 71

magnetic
– anomalies 89
– field 89, 90
– reversals 89
mantle 82
manufacturing process: iron and steel 11–12
Matthews, Drummond 89
medicines: species used to make 41–42
Mesosaurus 86
metal ores 95
metamorphic rocks 58, 70
metamorphism 78–79
meteors, meteorites 102
Milky Way 104
minerals 59
– in the soil 52
models
– astronomical 106–109
– of Universe's future 124
Mohorovicic discontinuity (Moho) 82
mudstone 79
National Park 49, 50
neutron star 118
obsidian 72
ocean ridge 87, 89, 91
ocean trench 88, 91
oil 67–68, 95
orbit 99
oxidation 12, 59
ozone 6, 7, 9–10
peridotite 83
photosynthesis 38
planets 99
plate boundaries 91
plate tectonics 91–95
polar front theory 25
Polaris, the Pole Star 96, 102
population
– factors affecting size 36
– growth, Egypt 51
power station
– coal fired 39
– hydroelectric 52
– solar 38
predator 33, 37
prey 33
principle of superposition; in rocks 68
pulsars 118
quadrat 36, 54
quartzite 60
radiation
– background (microwave) 122
– infrared 96
– ionising (from isotopes) 69
– microwave 121
– solar 38, 113–114
radioactive decay 69
radiometric dating 69
radiotelescope 98, 121
respiration 40
rain forest 42–43
relative dating 68
rift valley 77, 87
rivers 64

rhyolite 72, 73
rocks: recycling processes, 92–94
satellite 11
sandstone 60
San Andreas fault 93
sea floor spreading 87–90
schist 59, 93
sedimentary rocks 58, 66, 68, 70, 75
sediments 58, 60
seismic waves 80–84
seismometer 80–81
separation technique: fractional distillation 12
shadow zones 84
silicic 72
sill 71
slate 74
soil
– composition 71
– erosion 53
– formation 65
– profile 65
solar cell 113
solar energy 38
– detection at Earth's surface 113
solar system 99
– origin 111–112
space exploration 99
spectrum 98, 116
starvation and drought 53
stars
– life cycle of 117–118
– types 115–117
stomata 34
stratosphere 6, 7
stress 75
subduction zone 92, 93, 95
Sun: age and power of 113–114
supernova 118
super position 68
telescopes 96–97, 109, 119
transport, transportation: of rock particles 58, 61–64
troposphere 6, 7
traps: for capturing small animals 54
variable stars 116
viscocity 73
volcanos 71, 73–74
water
– as a solvent 21
– cycle 14
– shortage 16, 17–19
– structure of 20
wavelength: radiation from hot bodies 121
waves, seismic P, S and surface 80
weather
– forecasting 29–31
– maps 26, 31
weathering 58–60
Wegener, Alfred 85–86
wind, rock formation by 64
xerophytes 34, 35